New Innovations in Engineering Education and Naval Engineering

Edited by Nur Md. Sayeed Hassan and Sérgio António Neves Lousada

Published in London, United Kingdom

IntechOpen

Supporting open minds since 2005

New Innovations in Engineering Education and Naval Engineering
http://dx.doi.org/10.5772/intechopen.78503
Edited by Nur Md. Sayeed Hassan and Sérgio António Neves Lousada

Assistant to the Editor(s): Rafael Freitas Camacho

Contributors
Xiangxi Han, Jiaming Wu, Youhong Tang, Wei Lin, Zhiqiang Feng, Zhanbin Meng, Ang Qiu, Kant Kanyarusoke, Stig Ottosson, Hassan Karampour, Mahmoud Alrsai, Manuel Vicente Castillo Guilarte, Angela Bielefeldt, Marybeth Lima, Maria Lorna Azul Kunnath, Sérgio António Neves Lousada, João Pedro Gouveia, Rui Alexandre Castanho, Rafael Camacho

First published in London, United Kingdom, 2020 by IntechOpen
IntechOpen is the global imprint of INTECHOPEN LIMITED, registered in England and Wales, registration number: 11086078, 7th floor, 10 Lower Thames Street, London, EC3R 6AF, United Kingdom
Printed in Croatia

British Library Cataloguing-in-Publication Data
A catalogue record for this book is available from the British Library

Additional hard and PDF copies can be obtained from orders@intechopen.com

New Innovations in Engineering Education and Naval Engineering
Edited by Nur Md. Sayeed Hassan and Sérgio António Neves Lousada
p. cm.
Print ISBN 978-1-78984-037-7
Online ISBN 978-1-78984-038-4
eBook (PDF) ISBN 978-1-78985-560-9

For EU product safety concerns:
IN TECH d.o.o., Prolaz Marije Krucifikse Kozulić 3, 51000 Rijeka, Croatia, info@intechopen.com or visit our website at intechopen.com.

Meet the editors

Nur M. S. Hassan obtained his PhD in Engineering from Central Queensland University, Australia in 2011. He holds BSc degrees in Mechanical Engineering and Computer Science and is a recognized expert in computational fluid dynamics (CFD) at Central Queensland University. Dr Hassan has wide experience in the experimental study and numerical simulation of engineering problems, particularly relating to fluid flow systems, heat transfer, and renewable energy. He has published over 55 scientific articles in journals and conferences including two edited books and 7 book chapters. He is a reviewer of scientific articles of several journals and is a member of the Australasian Fluid Mechanics Society, the Australian Fluid and Thermal Engineering Society, the Mining Electrical and Mining Mechanical Society, the Australian Society of Rheology, and the Australasian Association of Engineering Education.

Sérgio António Neves Lousada is a Civil Engineer with a PhD in civil engineering and his area of knowledge is in hydraulics. He teaches at the University of Madeira, Faculty of Exact Sciences and Engineering, in the Civil Engineering course, with a major field of hydraulics, environment and water resources and secondary field of construction. Currently he is course director of the First Cycle in Civil Engineering, according to the statement of the Scientific Council of the Faculty of Exact Sciences and Engineering at the University of Madeira. Since 2012, he has been researching hydraulic sciences, particularly in the areas of hydraulics, urban hydraulics and marine and fluvial construction. He has published several articles and books and has participated in events mainly in the areas of hydraulics, urban planning, and land management.

Contents

Preface **XI**

Section 1
Engineering Design and Innovation Methods **1**

Chapter 1 **3**
How to Deal with Quantum Leap Innovations and Free-Fall Situations
by Stig Ottosson

Chapter 2 **17**
Service-Learning and Civic Engagement as the Basis for Engineering Design Education
by Angela R. Bielefeldt and Marybeth Lima

Chapter 3 **37**
Divergent Abduction Model and Its Convergent Interaction in Knowledge Production
by Manuel Vicente Castillo Guilarte

Chapter 4 **49**
Personalizing Course Design, Build and Delivery Using PLErify
by Maria Lorna A. Kunnath

Chapter 5 **67**
Mechanical Engineering Design: Going over the Analysis-Synthesis Mountain to Seed Creativity
by Kant Kanyarusoke

Section 2
Innovation Methods in Naval Engineering **89**

Chapter 6 **91**
Prologue: Exploring the Potential of the Sea
by Sérgio Lousada and Rafael Camacho

Chapter 7 **97**
Vortex-Induced Vibration of a Marine Riser: Numerical Simulation and Mechanism Understanding
by Xiangxi Han, Youhong Tang, Zhiqiang Feng, Zhanbin Meng, Ang Qiu, Wei Lin and Jiaming Wu

Chapter 8 **115**
Propagation Buckling of Subsea Pipelines and Pipe-in-Pipe Systems
by Hassan Karampour and Mahmoud Alrsai

Chapter 9 **139**
The Influence of Water Quality on the Structural Development of Vessels: Smart Dimensioning Process
by Sérgio António Neves Lousada, João Pedro Gouveia and Rui Alexandre Castanho

Preface

The demand for innovation in engineering is increasing. The new innovations are key elements of engineering education to provide an innovative engineering curriculum for the academics, researchers, and students with integrated and blended learning experiences. The objectives of this edited book are to provide guidelines in developing a successful engineering curriculum, in designing an innovative learning and teaching method, and in promoting consistent standards in engineering education.

This edited book will introduce resources and case studies of engineering in the areas of technology-enhanced engineering approach, integration of blended e-learning, and sustainable and social innovation in engineering learning experiences.

The book presents number of chapters focusing on the latest developments in engineering education and naval engineering and will provide the reader with a broad overview of the latest developments of engineering design and methods. The chapters include relevant technical, sustainable, and social innovations that have a significant influence on the society and the stakeholders. This edited book consists of two parts. Part One contains 5 chapters focusing on pedagogical aspects of engineering education, and Part Two comprises 4 chapters highlighting new innovations in naval engineering. A brief summary of each part is given below.

Part 1 Engineering and Innovations Methods

Chapter One *"How to deal with quantum leap innovations and free fall situations"* investigates three industrial case studies based on three different start situations for the development of the organization. These three situations are satisfying a market need; satisfying a want or a wish; and satisfying a creativity in combination with improvisation and the use of dynamic management and development methods. This study concludes an important aspect that the product and business development based on satisfying a *want* or a *wish* and mostly also a *need* does not follow linear or continuous curves. This makes the planning of activities difficult for practitioners. Therefore, people need to be creative, to improvise, and to use dynamic mind-settings and principles if they want to develop any successful innovations in their organization.

Chapter Two *"Service-Learning and civic engagement as the basis for engineering design education"* presents the pedagogy involved in service-learning that can be implemented to teach students the engineering design process. The similarities and differences of service-learning as implemented through engineering design are compared to community and civic engagement in disciplines such as social sciences. The study reveals the best practices in service-learning and civic engagement in engineering design produce better engineers and enables the profession to fulfill its highest purposes.

Chapter Three *"Divergent abduction model and its convergent interaction in knowledge production"* proposes a new approach that integrates the scientific method with the modified abductive method and its interaction with the inductive method based on divergence, establishing a new sequential form to search for and obtain alternatives in the generation of new theories by convergences, and in this way, gradually achieve a convergent procedure of support to produce new knowledge.

Chapter Four *"Personalizing course design, build and delivery using PLErify"* focuses on how a course building technique named PLErify was developed in response to emerging roles of faculty members and the need to implement their own teaching style in the technology driven teaching model as well as deep learning process. This study also explains an early attempt at pedagogy and trends that push the personalization movement for supporting the concept of personalized teaching and learning where technology application is most needed to make it more efficient, more effective, and more pragmatic.

Chapter Five *"Mechanical engineering design: going over the analysis-synthesis mountain to seed creativity"* demonstrates a change in delivering Mechanical Engineering Design (MED) to undergraduate students as an extension of Natural and Engineering Science. The study suggests a new teaching approach that attempts to align student thinking and learning activities with what exists in industrial MED. The new synthesis mode helps students build their self-confidence and better prepares them for industry.

Part 2 Innovations Methods in Naval Engineering

Chapter Six *"Exploring the Potential of the Sea"* presents the definition of naval engineering, which is the branch of engineering that has as the main activity the exploration of the potential of the sea. Although specialized, naval engineering is quite eclectic since it addresses the main aspects of other engineering modalities, directly or indirectly. With the development of oil exploration and production in the ocean, the naval engineer's work has extended to ocean engineering.

Chapter Seven *"Vortex-induced vibration of a marine riser: Numerical simulation and mechanism understanding"* focuses on Vortex Induced Vibration (VIV), which is the main cause of fatigue damage of the riser. The prediction of marine riser VIV is very difficult because of its strong non-linearity, instability, and uncertainty. In recent years, many numerical models of VIV for marine risers have been developed to explore the mechanism of marine riser VIV, providing scientific theoretical basis and practical engineering methods for vibration control and engineering design of marine risers. Combined with the authors' own recent research results, this chapter discusses the research progress and the key issues of marine riser VIV in ocean engineering, including phenomenon mechanism analysis and the development of different numerical research methods.

Chapter Eight *"Stability-of-subsea-pipelines and pipe-in-pipe-systems"* investigates buckle propagation of subsea single-walled pipeline and Pipe-In-Pipe (PIP) systems under hydrostatic pressure, using 2D analytical solutions, hyperbaric chamber tests, and 3D FE analyses. Experimental results are presented using hyperbaric chamber tests, and are compared with a modified analytical solution and with numerical results using finite element analysis for single-walled pipelines and PIPs.

Chapter Nine *"The Influence of Water Quality on the Structural Development of Vessels. Providing Directions for a Smart Dimensioning Process"* presents an extensive literature review articulated with practical approaches. This chapter aims to define relevant directions for vessel's structural development processes regarding the water quality (sea or river waters), where they will outline their routes. Therefore, the study looks for a relationship between the vessel's structural coating design process and the quality of the water where they navigate. Moreover, such a process not only will optimize/minimize the costs with the periodic maintenance of the vessel's linings, but also regarding its routes - contributing to the revitalization of their structural dimensioning.

We hope the selected chapters will help the readers to enhance their understanding of engineering pedagogy and processes.

Nur Md Sayeed Hassan
Central Queensland University,
Cairns Campus,
Queensland, Australia

Sérgio António Neves Lousada
UMa - Faculdade de Ciências Exatas e da Engenharia,
Campus Universitário da Penteada,
Funchal - Portugal

Section 1

Engineering Design and Innovation Methods

Chapter 1

How to Deal with Quantum Leap Innovations and Free-Fall Situations

Stig Ottosson

Abstract

When a quantum leap innovation enters the market, other competing companies can get in negative chaos and free fall situations for their businesses. To change such a situation, the use of controlled self-organization principles, for example, as prescribed by the Planetary Organization principles, can be valuable. This chapter, which is based on dynamic/flexible/agile philosophies and participatory action research from three industrial cases, mainly deals with the question how to develop quantum leap innovations and how to survive free fall situations. For both situations, creativity, improvisation, and dynamic development are essential ingredients for success. The chapter also gives some useful tips on how to successfully handle change management situations.

Keywords: innovation, dynamic principles, organization, planning methods

1. Introduction

"Innovation" has become almost a buzzword that has no one single definition of what it means. In general, it is a positively loaded term that brings hope in difficult times for actors in the private sector, the public sector, and the idealistic (nonprofit) sector as well as for whole economies.

However, to create successful innovations is complex and often influenced by unforeseen possibilities and situations as well as problems to overcome for which no known solutions exist. Even more unclear is how to develop "sustainable innovations," although we might have an intuitive feeling that the expression refers to developing something good for the society and its inhabitants.

This paper deals with some experiences from three practical innovation projects. All three cases were successful although the development was not as smooth as wanted or wished. The findings are highlighted in this paper.

2. Research

The paper is based on participatory action research (PAR) [1] from three industrial cases with three start situations: the development of SKF New Products (SNP) and the business unit FlexLink, the development of a new product of Careva system AB based on an invention, and the change management of Frontec Research &

Technology AB (FRT), which was a business in free fall. SNP was a subsidiary of the multinational ball-bearing company SKF AB. FRT was a subsidiary of Frontec AB, which was noted on the Swedish exchange market. Careva AB was a small Swedish privately owned company.

3. Theory

A practical definition of "innovations" is that they are new products (meaning goods and services) that are sold and used [2]. In this view, a new product or an *invention* is not an innovation but a part of it. "Sold" here means a wider view and money does not need to be exchanged. A new innovation can also be a new organizational method in business practices or workplace organization.

One often distinguishes between three main innovation types (*incremental innovations*, *radical innovations*, and *survival innovations* [2]), and the most common type of innovation seems to be the *incremental innovation*, meaning the continuous development of products and services as well as business models and also organizational settings until a disruption situation occurs (see **Figure 1**), often initiated by a competitor in the market. Lean product development is one method to successfully make continuous improvements of existing products and solutions. When a disruption situation occurs, there will be a free fall to a stop or a new opportunity to develop based on new knowledge or a new solution. This is sometimes called a *quantum leap* [3]. The free fall means a negative chaos, while a positive chaos and a euphoric feeling often occur when a completely new opportunity appears.

Note that the curves in **Figure 1** are not smooth in real development projects, which can also be seen from the cases described below.

Radical innovations appear as sudden steps up in the development level, sometimes called quantum leaps (see **Figure 1**). They are based on knowledge breakthroughs that are developed and introduced in the market. The technical development content is initially often large, and the organizational and business development comes after the introduction of the product on the market.

"Survival innovations" are needed for an organization when a radical change is needed for the business to survive. It can happen when a sudden step-down occurs (see **Figure 1**), for example, because of new laws or environmental changes, or when a market drop occurs for the business. This can be the result of competitors

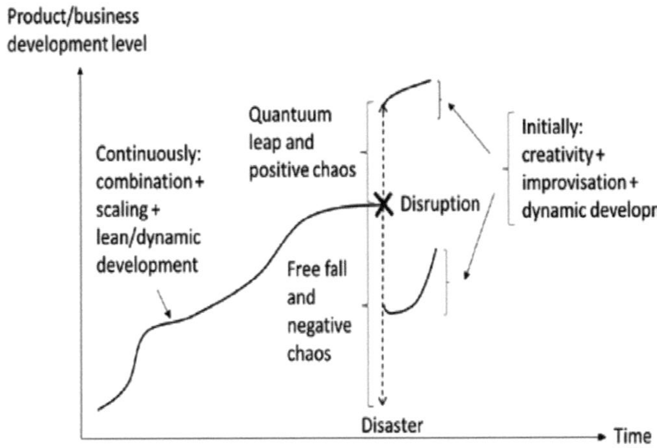

Figure 1.
Situational innovation (product and business) development [4].

introducing new technology and/or marketing and sales principles. In such a case, the technical development content is often initially small, and to improve the situation, focus must be on organizational and business development issues.

To coordinate all activities in the development of an innovation is a complex management activity for which some parts can be planned and for which *creativity* and *improvisation* are needed when it is not possible or meaningless to plan in depth.

"*Creativity* means shakings things up, both inside ourselves and in the world around us, and the constant re-organizing of both cognitive schemata and, to a greater or lesser extent, the domain of the creative person's activity" [4].

"*Improvisation* is thought of as making the best of things, while awaiting a return to the way things should be done. Improvisation is an exception, something we can 'fall back on' when things do not go the way they should" [5].

Therefore, innovation management is different from the management of mature businesses. Also, innovation management is place- and situation-dependent: the culture in the organization and the geographical area where the development takes place and the market where the initial marketing and sales takes place.

The more complex the leadership situation in a region or country, the higher the demands to produce market sustainable innovations and the more complex the innovation management as well as their costs will be. Sweden is an example of a country with such a demanding management situation (see **Figure 2**). Although Sweden's population is only equivalent to 0.13% of the total world population, it is the 33rd largest export economy in the world and the 4th most complex economy according to the Economic Complexity Index (ECI). Another example is that, although the Nordic countries, especially Norway, have high labor costs with an average of 125 than the G7 countries of 108, they have managed to be successful on a global scale. "Therefore, there are important lessons to learn from high-cost countries that successfully compete in the global marketplace" [6].

In principle, there are three main types of management situations: to develop incremental innovations in existing companies that are managed by *project leaders* and managers, to develop more radical innovations managed by *entrepreneurs* and *intrapreneurs*, and to develop "survival innovations" managed by *renovateurs* [7]. The three innovations appear in all sectors: the *private sector*, the *public sector*, or in the *idealistic/nonprofit sector*.

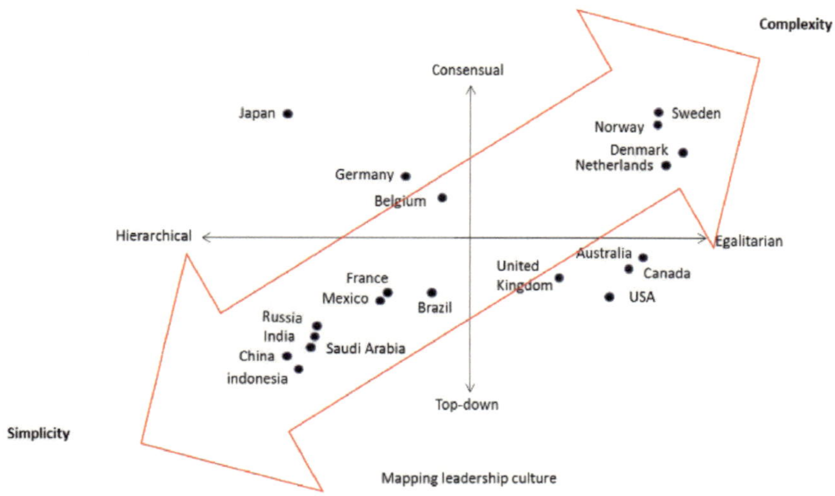

Figure 2.
Mapping leadership culture (based on [8]).

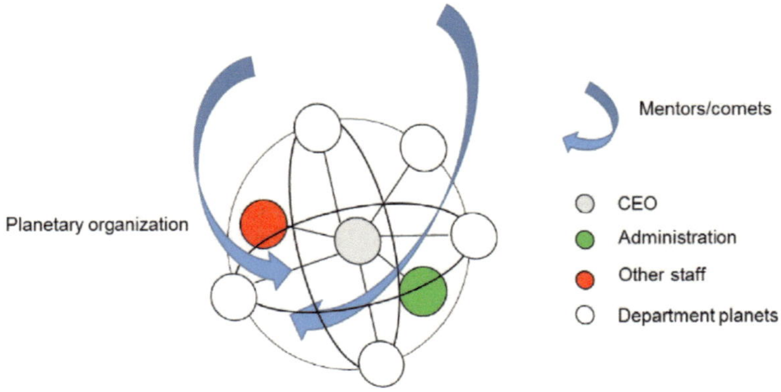

Figure 3.
Exercising dynamic leadership means to be present in the middle of the activities, allowing a controlled freedom in the team. In this case, the situation is pictured for a small- or medium-sized enterprise (SME) [4].

The classic view	The dynamic view
Small changes are negligible	Small changes can cause big differences
Linear—one best solution exists	Non-linear—many equally good solutions exist
Objectivity exists	The observer always influences
A small change in the initial conditions will not change the long-term behavior of a system	Just a small change in the initial conditions can drastically change the long-term behavior of a system
It is possible to make accurate long-term predictions about the behavior of the system	It is not possible to make accurate long-term predictions about the behavior of the system
Chaos is destructive	Chaos is the ground for development. Self-organization occurs out of chaos and disorder
A system is either stable or unstable	A system can swing between chaos and order
Organizations can be controlled/regulated	Organizations can only be influenced
One-dimensional up-down character (line organization with an order of importance)	Network character (Planetary Organization with more equality among its members)

Table 1.
Some examples of differences in view between the classic and dynamic views [2].

"Creative organizations should work out a model that would allow them to maintain a balance between non-hierarchical and hierarchical interactions, associated with equilibration and disequilibrium, or hierarchy, respectively" [9]. The Planetary Organization [10] seems to meet that demand well. This organization model can be seen as a combination of common line organization and self-organization. It was designed to take care of both vertical and horizontal communication and fast feedback, as shown in **Figure 3**. The leader is in the center of the Planetary Organization, giving energy to the other "planets." There are also "comets" which move freely in the Planetary Organization. The Planetary Organization has similarities with network organizations. However, they are more informal in their structures and are not thought to be something to use to actually organize an organization.

The Planetary Organization plays an important part of the Dynamic Product Development (DPD™) principles [2]. In turn, DPD™ is philosophically based on a dynamic view from quantum physics, chaos theory, and complexity theory, while the classic view—as Waterfall principles—are based on Newtonian mechanics, Taylorism, and the bureaucratic school. Some differences between the two

	Agile software development	Lean product development (LPD)	Dynamic product development (DPD™)
Background	Best practice	Best practice	Best practice + theory studies
Theoretical support	No theoretical foundation	No theoretical foundation	Quantum physics, chaos theory, complexity theory, innovation theory
Main research methods	Case studies (interviews)	Case studies (interviews)	Insider action research
Beneficiaries	Users, business	Customers, business	Users, business, and society
Leadership	No formal leaders	Management by wandering around (MBWA)	MBWA
Manning principles	Teams set up first	Teams set up first	Successive manning
Decision principles	Late final decisions	Late final decisions	Early preliminary and late final decisions
Location	Colocation	Colocation	Colocation
Work principles	- Iterations - Incremental steps - Frequent tests	- Iterations - Minimize waste - Quality assurance - Value streams	- Universal design - DFX order - Iterations within and between incremental work packages - Traffic light metaphor - Rules of thumb (e.g., BAD-PAD-MAD-CAD, flowing water principle, switch between topics, (e.g., to reduce waiting time, apply the Pareto principle), few demands to meet in each loop, and so forth)
Follow ups	Weekly meetings Performance, time (PT)	Weekly meetings Quality, cost, time (QCT)	Weekly meetings in the war room Performance, cost, time (PCT)
Comparing scientific studies with other PD methods	Nothing found	Nothing found	Yes, from hardware development, software development, and organization studies

Table 2.
Some important characteristics between three dynamic development methods.

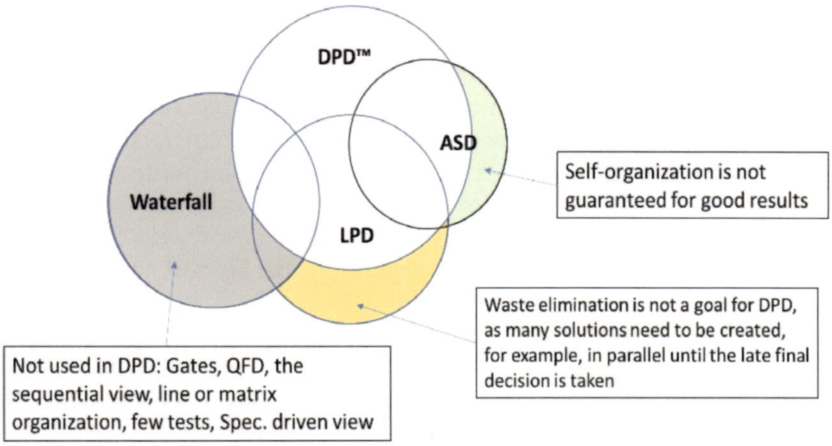

Figure 4.
Relations and differences between different development principles.

philosophic views are shown in **Table 1**. As shown, there are fundamental differences between the two views.

However, the dynamic view is gaining ground, for example, in Agile software development (ASD), and from there it starts to spread to hardware development (e.g., [11, 12]). Based on information, mainly from [2], comparisons can be made between different Agile/flexible/dynamic development methods (see **Table 2**). As seen in the table, there are many similarities between the different methods. DPD™ has more defined work principles and has a solid theoretical background, while the other methods are based only on best practice. **Figure 4** can be used to show the relationships and differences in a simple way. In that figure, the Waterfall principles included many principles that are also used in DPD™.

4. The three projects

4.1 Forming an incremental innovation

The Scandinavian division of the multinational SKF AB [13] decided to start SKF New Products in Gothenburg in 1979 as their volumes were decreasing mainly because of digital solutions. Therefore, a new start-up company was formed in their own premises in an old part of the SKF buildings, with its own board comprised of directors from SKF and two union leaders representing black- and white-collar worker unions. The author was recruited as an intrapreneur from another multinational company (ITT) for which he was responsible for the R&D laboratory in Stockholm. Before his appearance at SKF, 20 people from within SKF had been transferred to the new unit, mainly to work in one new business unit—SKF Actuators—and in an evaluation group of patents and product ideas. The treatment in this study is only about the evaluation group and what it accomplished.

One of the first decisions the intrapreneur made was to transform the evaluation group to the business unit, FlexLink, around an external prototype conveyor belt that the evaluation group was investigating. Having done that, the vice president of SKF AB told the intrapreneur and his staff directors that he had recently closed a business unit in Germany that worked with a similar product. On a direct question about whether he wanted the intrapreneur to also close this new business unit, he answered that it was up to the intrapreneur to decide on that. However, the staff directors ordered an external market investigation that showed a market and profitability situation far below the rules given for new businesses within the SKF Group.

However, the intrapreneur had experienced at ITT that market investigations for products that do not exist on the market are of limited value and that the entrepreneur/intrapreneur is of prime importance if a new product or service will be an innovation or not. The intrapreneur of SKF New Products had faith in the entrepreneur he had chosen for FlexLink and decided to act as an umbrella man sheltering his entrepreneur so that he was able to forget about company politics and to allocate all his energy into getting the new business unit running as fast as possible.

Next, there were many rules for the traditional business within SKF that had to be skipped, such as how business cards should look, the sizes of parcels for delivering the products, how marketing material had to be done, and pricing. Tricky questions included how the logo should look and if the logo should have the SKF mark or not and if FlexLink should have its own telephone lines. The intrapreneur of SKF New Products had learned that asking for permission was not the right way to proceed in the large ITT organization where employees simply did what they thought beneficial for the new business. The vice president of SKF, as well as different directors around him, did not consider it worth engaging in this small business that probably would have to be closed later as had happened with the German business.

Today "FlexLink is a leading conveyor manufacturer offering automated conveyor systems, flexible conveyor equipment, aluminum and stainless steel conveyors" [14]. According to their home page it has:

- 989 employees.
- 30% of employees are women.
- 50% of employees have a university degree.
- Operating units in 30 countries.
- Partner networks in more than 60 countries.
- More than 8000 installations worldwide, many for leading brands within FMCG, healthcare, automotive, and electronics.

Before FlexLink was purchased, partly by a Swedish venture capital company and later by the ABN AMRO Bank N.V. in the Netherlands, SKF made a considerable profit from running the business and the shares, although it took some years for them to break even. Had it been closed down early in its vulnerable development stage, which had been the natural decision based on the German experiences and the market investigation, SKF would have missed the benefits from the business unit.

4.2 Forming a radical innovation

In 2003, Mrs. Evastina Björk, who is an occupational therapist, defended her PhD thesis [15] in which she proposed a solution of a universal positioning belt for disabled people being transported in taxis and other vehicles. She had created that solution in 2002 and had made simple tests on it. Tailor-made positioning belts and modules of belt parts that could be combined to be suitable belts for people of different body shapes existed, but she found that was not a suitable situation for public transport, and a wish had been expressed by London Taxi to create a universal system. The new product concept consisted of two belts that were crossed over the chest of the passenger (see **Figure 5**).

In mid-2003, the development of the concept as a commercial product was started by the small company, Careva AB, in Gothenburg. The author was contracted for the technical development of the product, for which he used the development method Dynamic Product Development [2]. Quite soon, the functionality level reached about 80% of what was stipulated as the functioning level for the product (see **Figure 6**). DPD™ as a method was not commonly known, and classical methods, such as integrated product development and variants of it were commonly used. However, for financial reasons, the project had to be put on hold for almost 1 year (2004). When new funding was found, the money was coupled with demands for using classical development principles and classical control demands. A new start, orchestrated by an industrial design bureau, took place with a result that was not as good. From August 2007, the author was brought in again, and dynamic principles were again used to achieve a ready and functional prototype, which was achieved in February 2008.

Many small problems and tests on a large number of disabled passengers were needed to produce a commercial product and not until 2011 could the product be introduced on the market. From the homepage of Careva, the following information can be found:

"The Crossit was initially designed for public transportation such as taxis, busses, trains or aeroplanes, as the design allows someone who does not use positioning support but who wants to sit in that location in the vehicle to do so without removing the belt. It was designed to stay in place. However, after its introduction onto the market in 2011, the Crossit has also become popular for use in family cars.

Crossit has also been tested by users who employ a wheelchair, and it has been found to fit all wheelchairs with a sturdy backrest. It is especially suitable for elderly people who require posture assistance, helping them to achieve an upright sitting position, or for people who need extra support in a power chair. No tools are required for installation in a wheelchair or in a car seat. It does not cause any damage to interior of the vehicle."

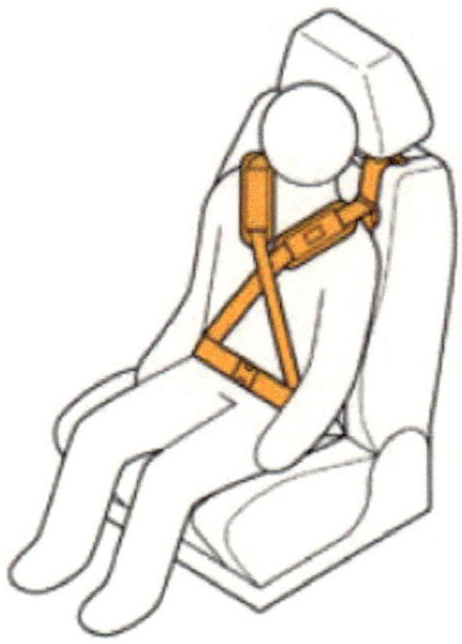

Figure 5.
The principle of the Careva Crossit positioning belt [16].

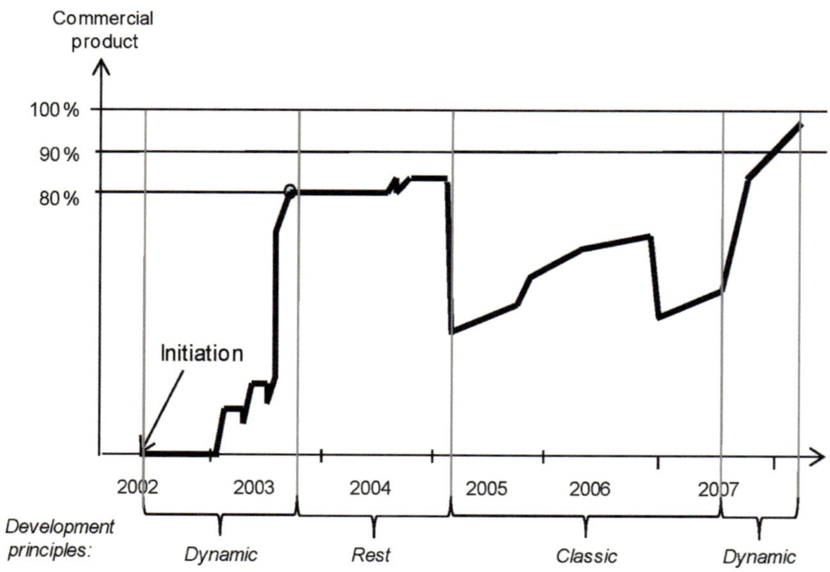

Figure 6.
The development of the Careva Crossit product [17].

To date, Careva exports the product to all the Nordic countries and several other countries around Europe.

4.3 Forming a survival innovation

A turnaround project was to bring Frontec Research & Technology AB (FTR) in Gothenburg, Sweden, to a profitable position within 6 months at an undefined cost payed by the business' own income and with support from the mother company, Frontec AB. Discussions between the chairman of the board and the author about such a project started in December 1999. Prior to that, the mother company had tried traditional ways to change the negative curves without success. The discussions ended with the author being given the mission, as interim CEO and renovateur, to turn around the company. If he should not succeed, it was determined that the company was to be closed down.

When the project started, FRT had 125 qualified employees. It made a loss of about 900.000 SEK (about 90.000 USD) per month and had done so for the last 6 months. When the negative trend had started about 1 year earlier, different experts were consulted to change the situation. In early December 1999, the CEO had decided to leave the company, so the mother company had to try something radical to change the situation.

The renovateur got a free hand to do what he thought best, without having to ask the board of the company for acceptance. Monthly financial reports had to be delivered to the mother company, which was noted on the Swedish exchange market. The renovateur started to plan his work in January 2000, although the formal start as deputy CEO was on the 15th of February 2000.

The change management/turnaround project started with the reshaping of the company to be a Planetary Organization, with the sun in the middle (the author/CEO), moons (the departments) in orbits around the planet, sub-moons (the groups) orbiting around each moon, and comets (free specialists) who moved freely in the Planetary Organization. Dynamic management principles and methods were introduced and used to speed up the changes needed to transform the company to a profitable enterprise (see [1]).

At the start of the project in February 2000, the company had no commercial value. Four months after the project start, the value of the company was calculated to be 32–44 MSEK or ca 3, 5–5 MUSD [18], and the company was sold in autumn 2000 to another company on the Swedish stock exchange—Sigma AB.

The combination of the Planetary Organization concept and the use of dynamic methods and principles were powerful in quickly getting the company to show blue figures, after its situation with heavy red financial figures. The activities specified below were especially important to improve their results:

- A "stripped-down" monthly financial report as a complement to the formal one for the mother company gave the CEO better possibilities to see how the core business was developing.

- An "early warning system" was introduced to take away unwanted surprises.

- The individual bonus system was changed for the managers to be a collective bonus system.

- Continuous improvements (lean) and the Pareto principle were used intensively.

- Management by MBWA was critical to pick up on weak signals and to improve the mood of the employees. MBWA also made it possible to meet the not confirmed bad rumors in a sophisticated way without revealing knowledge of the rumors.

- The three "comets" increased the efficiency of the work, based on storytelling and mentorship.

- By encouraging the consultants who were working at the premises of the customers to suggest small improvement projects and get feedback from the customers, the sales increased.

- The risk level for the company was lowered as each of the managers got to know the other managers' duties, activities, and situations meaning that they could better support each other and even take over their jobs in case of planned and unplanned situations.

A pure line organization was used before the change in management started. Very few of the actions carried out could have been accomplished without difficulty if self-management alone had been used. As there were people working against the changes while officially saying they were in favor of the new ways of organizing and working, they would have hampered success of the activities if pure self-management had been used. Based on these experiences, it is hard to see how self-management, a circular and/or a holacracy/holocentric/circular/sociocratic organization [19] could have been used to obtain the results achieved in this case.

5. Discussion

The three, with PAR [1] investigation, projects showed that the dynamic principles and especially the use of DPD™ gave successful results.

The first case—SKF New Products—shows some of the difficulties an intrapreneur has to face. In this case, the good thing was that the management of SKF did not believe in the business unit, FlexLink, so they did not care that the intrapreneur broke a number of company rules, such as forming a unique trademark and developing suitable marketing material for the business and its products.

The SKF case showed an innovation that was developed into a mature, sustainable, and growing business. The case shows that it was extremely vulnerable and needed to be taken care of in a very sensible way, getting shelter, care, and nutrition offered by the renovateur and using its own localities. **Figure 7** shows, in hindsight, the fruitful situation that was created for the large organization to take care of its lean and innovation activities living side-by-side. The innovation business in the figure is SKF New Products, with its different innovation projects organized as planets. The business units of SKF New Products remained in its own premises for all innovation projects until they had grown so much that they needed to move to new premises.

The SKF case also shows that market investigations for new products can be disastrous for deciding on whether further development should be done. A "good" entrepreneur will find his/her ways to success.

The second case—Careva—shows how tricky it can be to create to a commercial product and that product development can be far from the linear and nice looking S-shaped curves common in the literature.

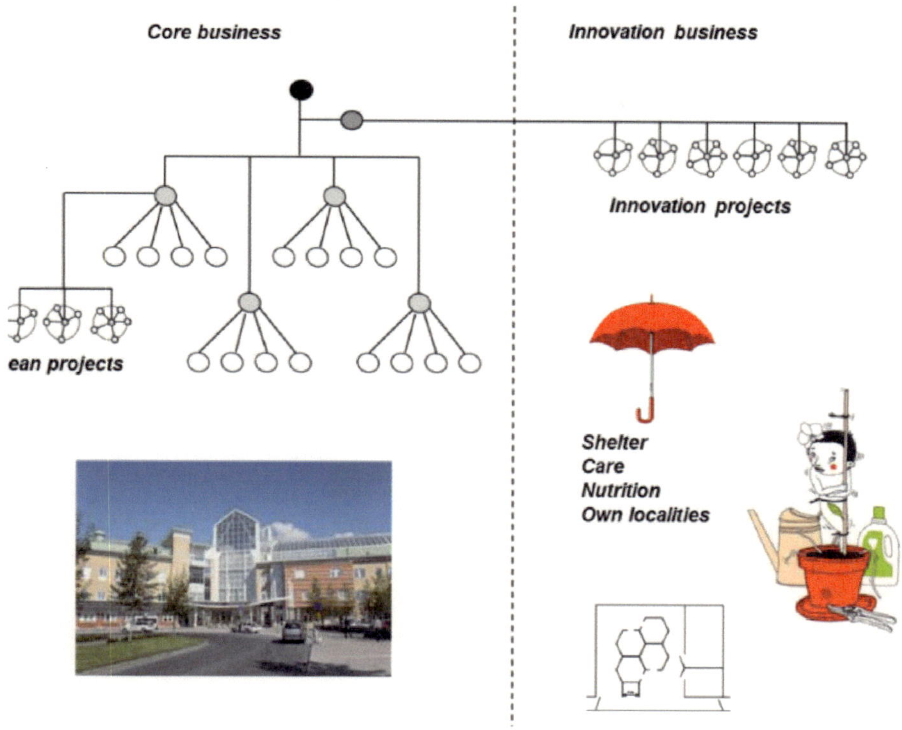

Figure 7.
Lean projects can be handled in the core business, while innovation projects is best taken care of outside the core business [2].

The third case—FRT—shows that Planetary Organization, in combination with dynamic methods, can give extraordinary results. Based on our experiences from the project, it can be concluded that neither the line or matrix organizations nor self-management organizations such as circular, holacracy, holocentric, or sociocratic are suited for handling modern, complex societal demands and fast-moving changes. The Planetary Organization accomplished this.

To determine whether the findings from the FRT case can be generalized, more projects need to be done and followed up. This is especially important, as traditional management methods and techniques are not designed to handle instability and rapidly changing situations.

6. Conclusion

The three principally different situations shown in **Figure 1** can be useful to have in mind when deciding what kind of organization, planning principles, and development methods should be used to get a wanted result in the end.

The three investigated innovation projects are examples of the fact that neither linear nor nonlinear curves can be used for the planning of new product development and innovation development activities. Often, only creativity and improvisation can push the development processes further when problems occur.

One conclusion is that product and business development based on satisfying a *need* and *a want* can be planned with traditional methods. Development based on satisfying a *wish* and a *want* can only partly be planned with traditional methods. In addition, creativity, improvisation, and the use of dynamic principles are needed to achieve successful innovations.

Entrepreneurship was shown, in these cases, to cover not being "politically right" and daring to take risks to break against accepted rules and opinions to be successful in the end.

All three development projects resulted in sustainable businesses.

Author details

Stig Ottosson
Norwegian University of Science and Technology, Gjøvik, Norway

*Address all correspondence to: stig.ottosson@ntnu.no

References

[1] Ottosson S. Participation action research—A key to improved knowledge of management. Technovation—The International Journal of Technological Innovation and Entrepreneurship. 2003;**23**:87-94

[2] Ottosson S. Developing and Managing Innovation in a Fast Changing and Complex World: Benefiting from Dynamic Principles. Vol. 2018. London: Springer Nature; 2018. ISBN 978-3-319-94045-8. 350 p

[3] Hamm S. Bangalore Tiger. New Delhi: Tata MacGraw-Hill; 2007

[4] Ottosson S. Three innovation development cases with three starting and performance conditions. In: ICBIA2018: 20th International Conference on Business Innovation and Applications; 20-21 August; Barcelona Spain. 2018

[5] Montuori A. The complexity of improvisation and the improvisation of complexity: Social science art and creativity. Human Relations. 2003;**56**(2):237

[6] Holtskog H, Carayabbus EG, Kaloudis A, Ringen G. Learning factories—The nordic model of manufacturing. In: Palgrave Studies in Democracy, Innovation, and Entrepreneurship for Growth. 2018. ISBN 978-3-319-41886-5

[7] Bjerke B. About Entrepreneurship. Cheltenham, United Kingdom: Edward Elgar Publishing; 2013. ISBN 978 1 78254 5385

[8] Meyer E. Being the Boss in Brussels, Boston, and Beijing. Harvard Business Review. 2017. Available from: https://hbr.org/2017/07/being-the-boss-in-brussels-boston-and-beijing, Retrieved October 2019

[9] Shkliarevsky G. Understanding the process of creation: A new approach. Journal of Sustainable Business and Management Solutions in Emerging Economies. 2017;**22**(3). DOI: 10.7595/management.fon.2017.0021

[10] Ottosson S. Planetary organizations. Technovation. 1998;**19**(2):81-86

[11] Schmidt TS, Paetzold K. Agilität als Alterative zu traditionellen Standards in der Etwicklung physischer Produkte: Chancen und Herausforderungen (in German). In: DfX-Symposium 2016; 05-06 October; Jesteburg, Germany. 2016

[12] Böhmer A, Grauvogl C, Schwiegert-Recksiek S, Becerril L, Bahrouni Z, Lindemann U. Towards agile development of physical products. In: 23rd ICE/IEEE ITMC, Madeira, June. 2017

[13] Available from: www.skf.se [Accessed: 14/03/2018]

[14] Available from: www.flexlink.com [Accessed: 14/03/2018]

[15] Björk E. Insider action research applied on development of assistive products [PhD thesis]. Magdeburg, Germany: Otto-von-Guericke-University; 2003

[16] Available from: www.careva.se [Accessed: 14/03/2018]

[17] Björk E, Ottosson S. Lessons learned from a want based NPD project. In: International Design Conference—Design 2008; May 19-22; Dubrovnik, Croatia. 2008

[18] Markkula, Engstrom. Företagsvärdering av Frontec Research & Technology AB (in Swedish). Retrieved from report to the board of

Frontec AB from SET-Konsulter AB in Stockholm. 2000

[19] Ottosson S, Björk E. Experiences with planetary organization. In: The International Conference on Business Case Research (ICBCR 2016); December 12-14, 2016; Hong Kong. 2016

Chapter 2

Service-Learning and Civic Engagement as the Basis for Engineering Design Education

Angela R. Bielefeldt and Marybeth Lima

Abstract

Service-learning (SL) is among the pedagogies that can be used to teach students the engineering design process. The similarities and differences of SL as implemented via engineering design are compared to community and civic engagement typical in disciplines such as social sciences. Although engineering design can be conceptualized via a number of paradigms, a human-centered design approach is particularly well-suited to SL projects. SL projects typically engage engineering students and instructors with stakeholders who do not have technical backgrounds. This approach is different than many industrially-sponsored projects that are more typical in capstone design projects and poses unique challenges and opportunities for engineering design education. Best practice recommendations for SL design projects have been distilled, with a particular emphasis on developing reciprocal partnerships and meaningful student reflection. SL design projects can lead to a rich array of knowledge, skills, and attitude outcomes among students, including ethical development, humility and empathy, and creativity and innovation. Enhanced recruiting and retention using this pedagogy has also been reported. Assessment of community partner satisfaction, learning, and outcomes are generally less well documented. SL design projects can be integrated into courses ranging from first-year to senior capstone, providing benefits to communities while enhancing students' skills.

Keywords: service-learning, community engagement, human-centered design, reflection, reciprocal partnerships, professional skills, attitudes, empathy, humility

1. Introduction

In the twenty-first century, "engineers are called to be change-makers, peacemakers, social entrepreneurs, and facilitators of sustainable human development" [1]. Preparing engineers to meet these challenges requires a rich educational experience. In particular, the way in which students are taught the design process is important. The products, processes, and infrastructure designed by engineers are critical to human quality of life, with an array of positive and negative impacts that should be carefully considered. More broadly, the designs of engineers are having global environmental effects. A rich design experience will reinforce to students the coupled socio-technical challenges they will face in practice, and prepare them to recognize and wrestle with the complex array of ethical issues that are inherent in all designs.

It is not sufficient that engineers have a great depth of technical knowledge, so-called I-Type education. Engineering education has been moving toward a T-shaped model that adds breadth skills that cross the boundaries of a single profession, such as teamwork, communication, and global understanding [2, 3]. Perhaps we need to move beyond T-shaped engineers to envision "cluster" type engineers [1], who will sit with a broad array of stakeholders (including members of the public and those in policy, social scientists, and natural scientists) to design appropriate and sustainable processes and products that better meet an array of environmental, social, and economic objectives.

It is our claim that service-learning can serve as an ideal basis for design education that strives to meet the aforementioned goals of educating global citizen engineers. In addition, the hard work invested by students and educators can yield tangible results that serve real people, as opposed to designs in AutoCAD or objects that are displayed at a design fair and then go to waste. Engaging with communities may also broaden the diversity of students interested in becoming engineers, both in terms of recruiting students into engineering majors in higher education as well as retaining students to graduate with engineering degrees and enter the engineering workforce [4].

This chapter begins by defining service-learning (SL) and community engagement and briefly describing their history in higher education and in engineering. Next, frameworks and theories of design that are particularly relevant to SL are presented, with a focus on human-centered design. This section is followed by a discussion of essential elements of SL-based design projects, as well as challenges and pitfalls of SL as a pedagogy for design education. The student knowledge, skills, attitudes, and identity that can result from SL-based design projects are presented next. Examples of SL-based design programs and courses are integrated throughout the chapter to illustrate concepts and best practices. This chapter is intended to provide the reader with an introduction to service-learning as a vehicle for design education, and to provide additional resources for readers who wish to delve into more detail with the theory and practice of this pedagogy.

2. Service-learning in engineering education

Service-learning is defined as "a credit-bearing, educational experience in which students participate in an organized service activity that meets identified community needs and reflect on the service activity in such a way as to gain further understanding of course content, a broader appreciation of the discipline, and an enhanced sense of civic responsibility." [5] Service-learning in higher education was pioneered by Ernest Boyer [6, 7] and other scholars in non-engineering professions [8–10] and was identified by George Kuh [11] as a high impact educational practice critical to the retention of early career college students. Service-learning, and more broadly civic engagement, which encompasses curricular and co-curricular efforts to ensure that the university is using its resources to partner with communities and other stakeholders to address complex societal issues, are a well-defined part of the higher education landscape in the USA. Campus Compact, the major professional society for civic engagement in higher education, has more than 1100 universities as members.

Models of service-learning were presented by Heffernan [12], and include (among others) a discipline or placement based model, in which students are situated within the community and perform community service to meet their learning objectives, as well as a problem-based or deliverable model, in which student create or co-create (with community) a product to fulfill course requirements. Service-learning in engineering has largely used the deliverable model, in which students deliver designs or designed and built artifacts.

Leah Jamieson pioneered service-learning in engineering through the Engineering Projects in Community Service (EPICS) program at Purdue University [13, 14]. This model features vertically integrated teams consisting of an approximately equal number of first-year, sophomore, junior, and senior engineering students who take a course repeating times for semester credit and who work together on addressing community issues using human-centered design. The teams are also multidisciplinary, including students studying an array of engineering and non-engineering disciplines. The community partnerships are often long-standing, with EPICS conducting a number of projects with partners over many years. Examples of projects conducted by EPICS in partnership with communities include hands-on exhibits for science museums, custom toys for children with disabilities, and software for elementary schools, non-profits, and public agencies. The EPICS model has expanded to include approximately 40 colleges of engineering nationally and internationally [15]. Edmund Tsang [16] is the editor of the engineering volume of the American Association of Higher Education's Service-Learning in the Disciplines. Numerous early models of service-learning in engineering are shared in this volume.

Though there is much work on service-learning in engineering, engineers serving the common good through co-curricular (outside the classroom) methods also play a large role in learning through service (LTS) activities [17, 18]. Many pre-professional and practicing engineers have participated in engineers without borders (EWB), whose mission is "To be the beating heart of the engineering movement for sustainable global development, building and evolving engineering capacity throughout the world." (http://ewb-international.com/). In this context, engineers partner with communities throughout the world that have a lack of access to resources in an effort to improve the quality of life for people in these communities. Common projects include improved sanitary conditions, enhancing water quality and availability, and access to energy.

There has been a proliferation of curricular and co-curricular opportunities for civic engagement in engineering since the turn of the century. SL design projects have been integrated into introductory courses for first-year students, technical core courses, and senior capstone design. Readers are encouraged to consult the *International Journal for Service Learning in Engineering: Humanitarian Engineering and Social Entrepreneurship* (IJSLE), especially two special issues published in 2014 and 2015, Opportunities and Barriers to Integrating Service-Learning into Engineering Education [19] and University Engineering Programs that Impact Communities: Critical Analyses and Reflection [20]. Additionally, the Community Engagement Division of the American Society for Engineering Education was created in 2012 and has a resource page for general knowledge in this area (https://aseeced.libraries.psu.edu/resources).

3. Design frameworks

The design process can be modeled in a number of ways, with specifics that vary somewhat depending on whether engineers are designing infrastructure at the community scale (e.g. a bridge, road, power system), physical products that are owned at a household or personal level (e.g. a car, computer), or processes (e.g. computer software). Some methodologies are more congruent than others with service-learning. The human-centered design process has often been used to frame service-learning (e.g. [21, 22]), and also aligns with numerous elements in the conceive-design-implement-operate (CDIO) process [23]. Human-centered design puts the people who are the users/community members at the heart of the process, engaging them throughout all phases. Optimally, service-learning embraces the notion of designing with communities. **Figure 1** offers a visual representation of the human-centered design process. The hexagon in the center represents the team of people working together on a particular issue (inspired by [1]), which is embedded in the complex ecosystem of the technical, social, and environmental realms. The community members (C) are "at the table" working side-by-side with engineers (E) and other experts in policy (P) and natural and/or social scientists (S). There are opportunities to harness community expertise in all phases of the design process.

An individual or the community collectively should identify a problem or situation they believe engineers might be able to contribute to solving or improving. The community should be the driving force, with a vision of partnering with engineers. In other words, problem identification should not be externally imposed. An engineer might share data with the community that she/he believes indicates an issue, but should not presume that her/his external perceptions of a 'problem' are authentic to a specific individual or community. Otherwise, there is an implication that a particular community or individual is at a 'deficit', needing charity or help from an "expert" engineering student, versus being co-equal partners in working to improve a situation.

Once an issue has been identified by the community, the next step is to gain a thorough understanding of the issue. It is important to realize that a particular problem is situated within a larger framework of the planet and environment at large, the society and economy in which a community or individual resides, various cultural norms and legal constraints, and interactions among these complex systems. Engineers should

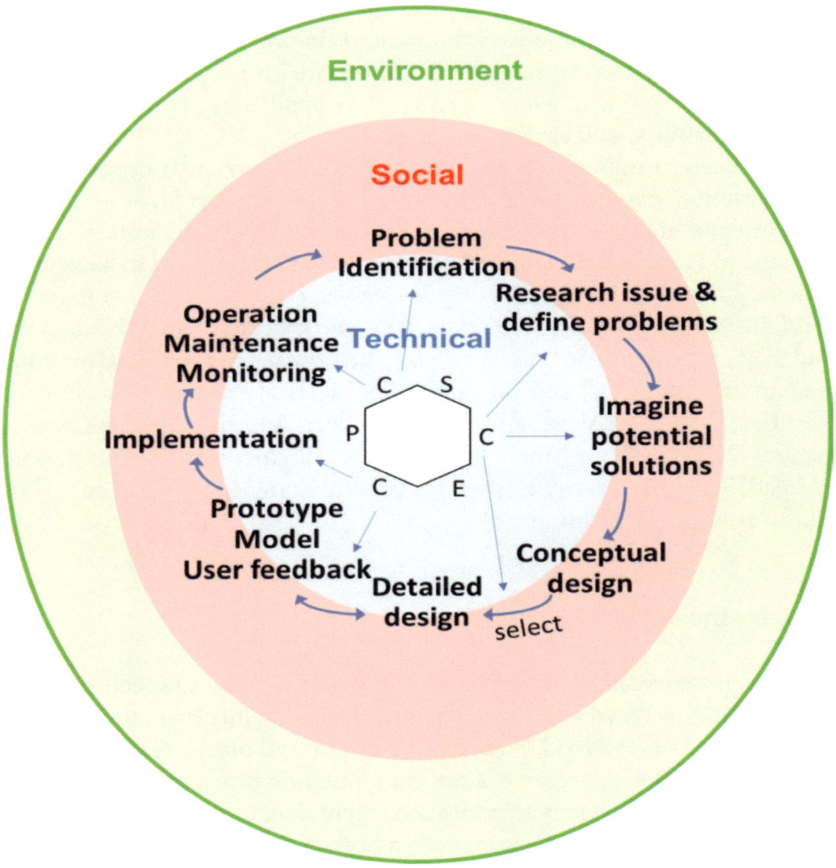

Figure 1.
Conceptual model of the human-centered design process as a collaboration among engineers (E) and community members (C) with contributions by policy makers (P) and scientists (S), situated within larger environmental, social, and technical realms.

have a strong understanding of the technical issues that are relevant to a problem, as well as community issues that they can gain perspective on through research. Critically, they also need to partner with others "on the ground" to fully understand other conditions relevant to the problem. In this stage, students should talk with and listen to their community partners. Ideally, this process includes contextual or transformational listening, which is a skill that must be thoughtfully developed [24–26]. The public and community should not be viewed as a monolith; there are sure to be an array of individuals and groups with different perspectives. Engaging an array of stakeholders early in the process can yield important benefits. The more students in their role as novice engineers can immerse themselves in the communities and with the people their engineering is designed to serve, the more likely they are to better understand and appreciate the needs of the ultimate users of the co-created design. This approach aligns with the ideas of empathic design [21, 27]. Students may also need to recruit partners or work with other disciplines to gain a thorough understanding of relevant constraints and criteria.

The next phase in the process focuses on divergent thinking, where individuals imagine an array of potential solutions. Engineers often bring examples of solutions that have worked in similar situations. But each situation is unique, and engineers should not force fit technology to a problem. The analogy is often that engineers have a set of tools, and just because they have a "hammer" does not mean that is the right tool for the job. Students should not position themselves in roles as experts, but as learners, collaborators, and facilitators, bringing their ideas and inviting ideas from others. Interactive discussions with a broad array of stakeholders are likely to yield a diverse array of creative ideas. This step is critical to the process, in order for the best solutions to be among the array of options being considered.

Next, there should be a thoughtful process of evaluating the range of ideas under the set of local constraints and criteria, to narrow in on a sub-set of potentially feasible, appropriate, and optimal solutions. This process should be conducted by the community members and engineering students working together in a participatory design process. The evaluation process should consider the larger context of the issue, including the social and environmental spheres. Engineers then create conceptual designs, which allow rough evaluation of metrics such as cost, environmental emissions, etc. Typically a number of the important criteria that determine an optimal solution are subjective. Thus, community members must be engaged in contributing to the design and evaluating these issues. The community should select the 'optimal' solution from among the sub-set of options that went through the conceptual design phase. This is a convergent phase of the design cycle, and may be challenging given that different stakeholders may have different perspectives on 'optimal.'

Engineers then typically handle the majority of the detailed design phase, which largely resides in the technical realm. Engineering students may complete this work if carefully supervised by instructors with appropriate expertise; some projects will require that licensed Professional Engineers review the designs. More forward-thinking SL programs are engaging in co-design among community members, students, and engineers. Where appropriate, prototypes of products are created, which can then go through testing by the community. In the case of infrastructure, computer models are built and subjected to expected human and natural conditions (e.g. hurricane); results are shared with stakeholders. Design changes can be made in response to the testing feedback cycle. This iterative process can often be viewed as a microcosm of the full design process (e.g. a problem might be identified in the prototype, alternative fixes are proposed and evaluated, etc.). The teams of engineering students and faculty should be completely transparent with stakeholders, explaining what they are doing and why. This approach provides an opportunity for co-equal learning among all of the participants in the design process, and is inclusive of both community members and engineering students.

The implementation steps, such as manufacturing a designed product, are often thought of as 'detached' from users and communities. However, in service-learning projects there are often opportunities to engage communities in this phase. For example, community participation in constructing a school playground, building a Habitat for Humanity home, community participation in building a Bridges to Prosperity (B2P) bridge, and locals producing ceramic water filters for point-of-use household treatment of drinking water in a micro-enterprise [19, 20]. Community involvement in the implementation step can be particularly impactful and contributes to the community "taking ownership" of the constructed artifact that they co-designed and helped to construct. The same is true in the operation, maintenance, and monitoring phases of a project. Community understanding of the process and ultimately their sense of ownership is fostered by their intimate involvement in all phases. The greater the participation of the community in all phases of the project, the greater the overall sustainability of a project over the long term—and across the interconnected areas of societal, environmental, and economic issues.

Done well, service-learning enacted through a model of human-centered design requires frequent engagement with the community across all stages of the design process. The more engaged community members are in the entirety of the design process, the better the outcome will fulfill project goals. Community members may not be immediately available at the discretion of a student design team, and communication processes and timelines need to be respectful of these preferences and needs. The feedback cycle among members of a design team that stretches across disciplines requires thoughtful consideration at each step. Catalano [28] advocates for a contemplative paradigm, which he combined with service-learning in a senior capstone design course. The various elements in the human-centered design process imply that a majority of significant service-learning design projects will have timelines that stretch beyond the confines of a single academic term. This "feature of the landscape" requires creative thinking to integrate community-scale design problems into higher education, adapting traditional course structures (e.g. [29] 'tyranny of the semester'). A thoughtful process to design the SL experience is encouraged. The Learning Though Service Program Model Blueprint is a tool that can facilitate this process, considering the perspectives of a wide range of stakeholders (e.g. students, community members, instructors, the university, intermediaries such as non-governmental organizations, practitioners) with respect to value propositions, relationships, and resources [30].

A sub-set of engineering service-learning design focuses on poverty alleviation, in programs such as Humanitarian Engineering and Engineering for Developing Communities. Nelson [31] described four different mental models that are commonly used to frame design processes associated with poverty alleviation: income first, needs first, rights first (including human-centered design), and local first. A well-being framework brings these four mental models together. The framework supports the importance of deeply engaging with communities and recognizing their unique expertise in their local context. Because poverty is framed as "the systematic failure to achieve wellbeing objectives", the framework lends itself to a series of metrics that can form the basis of design objectives, constraints, and criteria; for example, "material sufficiency, bodily health, social connectedness, security, and freedom to make choices around action" (p. 2). A service-learning design program at Ohio Northern University is a case example of the well-being framework [31].

Entering into service-learning design projects, instructors may want to consider servant-leadership as a framework for their teaching and as a model for students to consider when they engage with communities [32]. Design instructors will have a role as a "guide on the side", with a mindset of mentoring or serving both their students and the community partner, and being mentored and served by these constituents. A case study of this approach was a service-learning project in a senior thermodynamics course at the Milwaukee School of Engineering [32]. The LSU Community Playground Project, which is affiliated with a first-year engineering design course, required the service-learning instructor to develop a servant leadership approach to be successful; the evolution from becoming a "traditional" engineering educator to a servant leader engineering educator is described in [33]. Stoecker [34] takes this concept further, suggesting that engaged faculty frame their work as community organizing.

4. Essential elements and challenges of SL-based design projects

There are several essential elements of successful service-learning-based projects. The authors strongly suggest that faculty who wish to use this pedagogy work with their university's office of civic engagement and/or service-learning to help identify community partners and to assist with planning and executing their projects within a reciprocal framework. Other groups, such as non-governmental organizations (NGOs), may be key stakeholders, particularly in international service-learning projects.

In terms of reciprocal partnerships, an asset based model of collaboration is ideal because it acknowledges the resources and assets that the university and community "bring to the table," as well as identifies the needs that each constituent seeks to meet through partnership. For example, universities might have assets with respect to discipline-specific knowledge and monetary resources, while communities might have assets with respect to community-specific knowledge and capacity resources. Partnerships are more successful when constituents combine their strengths to address a community issue together rather than a charity model in which one constituent helps the other. Another way to frame this asset based philosophy is that each constituent will both learn something from and teach something to the other.

The 2006 Community Partner Summit [35], p. 13 and Portland State University's 2008 Partnership Forum [36], p. 3 identified the following essential components for successful community-university partnerships:

1. Quality processes (open, honest, respectful; relationship-focused, characterized by integrity; trust-building; acknowledgement of history, commitment to learning and sharing credit)

2. Meaningful outcomes (specific and significant to all partners)

3. Transformation (at individual, institutional, organizational, and societal levels)

These essential components are achieved by practicing the following processes ([36], pp. 3–4):

- Asset (resources, strengths, and interests) identification and recognition for all partners

- Dialog within partners and between partners

- Creation of common language

- Relationship-building strategies
- Describing and understanding each other's culture
- Learning together
- Collaborative problem posing and solving
- Collaborative agenda setting
- Identification and recognition of each partner's needs, issues and challenges
- Self-assessment and reflection within each partner group and between partners
- Constant negotiation and modification
- Supporting infrastructure in each partner's organization

Another important component of a successful service-learning partnership is reflection, or metacognition. Professionals constantly reflect on what they are doing, why they are doing it, and next steps; students need to develop this skill that professionals may forget that they practice, because this practice is so embedded in their daily work. There are many models of reflection ranging from the simplest (what, so what, now what) to those that are more complex [37, 38]. Lima and Oakes [39] have a list of reflection questions in Chapter 2 of their textbook on service-learning in engineering. Reflection can be used to catalyze and assess student learning.

A thoughtful assessment plan should be developed, to help ensure that the outcomes desired for both communities and students are achieved. This plan should include formative assessment to enable during-course adjustments, as well as summative assessment to provide 'lessons learned' for the future. Assessment methods for student outcomes are well documented (see examples in [40]). Community outcomes have been rigorously studied in fewer instances, and are an area where additional scholarship is needed.

Even when adhering to all essential components and processes for successful partnerships, there can still be challenges and pitfalls. For example, as mentioned previously, it can be difficult to manage partnerships within the time constraints of a semester: most community issues involve people working on them throughout the year, not in 15-week blocks. This constraint may require some thought in terms of deploying a design and maintaining it once it is built. Repeating courses with the same community partner is one way to address this issue; others have created infrastructure to complete and maintain projects [39, 41]. Such considerations ensure that a design effectively serves the community, instead of being dumped on the community. Student resistance to participating in service-learning classes is also possible [32]; explicitly and repeatedly connecting the service activities to the learning objectives in class allays most student concerns. Finally, communication can be an issue, particularly where media is concerned. University media tend to focus on the students and faculty involved in a service-learning project and typically portray the community-university relationships as the university helping the community [42]. An explicit conversation among constituents about uniform talking points for media, and if at all possible, media interaction with all constituents present, is recommended. See [42], for more details.

5. Potential student outcomes of SL-based design

Across all disciplines, service-learning has been shown to be an impactful pedagogy. A recent meta-analysis of SL across 62 studies (all included a control group, elementary through postsecondary level students with 68% college undergraduates) determined that SL resulted in "significant gains in five outcome areas": academic achievement (grades or test performance; highest mean effect size, ES, 0.43), social skills (leadership, cultural competence, social problem solving; ES 0.30), attitudes toward self (self-esteem, self-efficacy, personal abilities, feelings of control; ES 0.28), attitudes toward school and learning (academic engagement, enjoyment of course; ES 0.28), and civic engagement (civic responsibility, altruism; 0.27) [43]. It is unclear whether or not any of the studies included in the meta-analysis included engineering students, but the results are nevertheless compelling.

Within engineering, previous research has identified a number of knowledge, skills, attitudes, and identity (KSAI) outcomes that could result from engineering student engagement in project-based service-learning (PBSL); [40] presented a literature review from numerous published sources. While that study extended beyond SL in design settings, SL-based design should have the capacity to yield the same array of outcomes. SL-based engineering design education can achieve all of the core technical outcomes one would expect from engineering design in general (aligned with the academic achievement outcome in the meta study), while also realizing a number of additional outcomes. The potential outcomes of SL-based design education that map to the technical and professional knowledge and skills expected of engineers internationally and by U.S. accreditation are summarized in **Table 1** [44, 45].

A greater complexity and range of design constraints are typical in SL-based projects compared to other design experiences. Service-learning executed through human-centered design may be superior to standard design pedagogy in developing

IEA Washington Accord Program / Graduate [44]	ABET Engineering Accreditation Criteria [45]
Complex engineering problems: WP3 No obvious solution and require abstract thinking, originality in analysis.... WP5 Outside problems encompassed by standards and codes.... WP6 Involve **diverse** groups of stakeholders with widely varying needs	Complex engineering problems... no obvious solution, ...not encompassed by current standards and codes, involving diverse groups of stakeholders, ...
WK7 Comprehend the **role of engineering in society**... ethics and the professional responsibility of an engineer... impacts of engineering activity...	
WA3 **Design** solutions... that meet specified needs with appropriate consideration for public health and safety, cultural, societal, and environmental considerations	2. Apply engineering design... with consideration of public health, safety, and welfare, ... global, cultural, social, environmental, and economic factors
WA6 Apply reasoning informed by contextual knowledge to assess societal, health, safety, legal and cultural issues and the consequent responsibilities relevant to professional engineering practice...	4. Recognize **ethical** and professional responsibilities in engineering situations and make informed judgments, which must consider... global, economic, environmental, and societal contexts
WA7 Understand and evaluate the **sustainability** and impact of professional engineering work...	
WA8 Apply **ethic**al principles...	
WA9 Function effectively...as a member or **leader** in diverse **teams** in multi-disciplinary settings	5. Ability to function effectively on a **team**... provide **leadership**...
WA10 **Communicate** effectively on complex engineering activities with the engineering community and with society at large....	3. Ability to communicate effectively with a range of audiences
WA12 ...have the preparation and ability to engage in independent and **life-long learning**...	7. Ability to acquire and apply new knowledge as needed...

Table 1.
Knowledge and skill outcomes achievable via SL-based design and PBSL.

communication skills with diverse audiences and teamwork/leadership skills in interdisciplinary settings. In addition, PBSL in engineering has been shown to yield enhanced creative design; cultural competency and leadership (social skills); self-confidence; attitudes toward community service; and engineering identity. The compiled data in [40] indicated outcomes for which the projects with a SL context yielded enhanced outcomes in comparison to non-SL projects.

SL-based design embeds an array of ethical issues, both microethics and macroethics, and may be particularly impactful in building students' ethical reasoning skills. In a faculty survey on ethics and societal impacts instruction, 212 respondents who described their capstone design course as including ethics and/or societal impact topics indicated that these topics were taught via service-learning [46]. Zoltowski and her collaborators [47] have been developing instruments and methods to measure ethical gains as a result of SL-based design experiences (e.g. [48]).

In addition to knowledge and skills, attitudes are important to the professional success of engineers and are explicitly recognized in CDIO [23] and the American Society of Civil Engineers (ASCE) *Civil Engineering Body of Knowledge for the 21st Century* (CEBOK). The third edition of the CEBOK [49] explicitly includes affective domain goals and rubrics associated with seven outcomes. Attitudes supportive of professional practice that may be specifically developed via a SL design experience, such as "value effective and persuasive communication to technical and non-technical audiences" which requires "empathy... with diverse clients and stakeholders" ([49], pp. 2–42–43). The professional attitudes listed in the CEBOK3 (pp. 2–53) and developed specifically via SL may include creativity, flexibility, consideration of others, empathy, honesty, integrity, respect, sensitivity, thoughtfulness and tolerance. Humility [50] and empathy [51] have been proposed as important mindsets in working with communities.

Of additional interest is the extent to which SL-based design is effective at developing students' creativity and innovation skills. This has not yet been rigorously studied using established instruments (such as the Creative Engineering Design Assessment Purdue Creativity Test or Purdue Creativity Test [52]); rather, the data reflects student self-assessments in surveys or anecdotal statements by instructors. One of the more rigorous assessments was associated with a first-year mechanical engineering design course [53]. A sub-set of the design projects were SL-based and included leadership training. Students engaged in SL projects had a statistically significant gain in the self-assessed extent to which they possessed creativity/ingenuity on the post- versus pre-assessment using a five-point scale; gains were not statistically significant among students working on non-SL design projects. In a senior product design course with service-based projects, students rated their creativity at a higher level on the post-survey than the pre-survey (average ~6.55 increased to ~6.95 on nine-point scale; $p < 0.05$); this compared to a gain of about one-point in their self-rated product design skills [54]. Fully anecdotal statements regarding growth in students' creativity and/or innovation skills in association with service-based design projects were made in a number of other papers [55–61].

Another set of proposed outcomes from SL-based design is that it may help attract students to engineering majors and/or retain students in engineering, particularly women and underrepresented minorities. Many students are drawn to engineering due to a desire to make a difference, help others, and improve society. SL projects offer tangible examples of these outcomes, inspiring students and providing rewarding experiences. Three large service-learning programs in engineering have data related to the impacts of their program in recruiting/retaining female students: the Service Learning Integrated Throughout a College of Engineering (SLICE) program at the University of Massachusetts Lowell [62], EPICS at Purdue University [63], and the Humanitarian Engineering and Social Entrepreneurship (HESE) program at Pennsylvania State University [64]. Other SL programs have reported on the large percentage of women among the participants, such as the Humanitarian Engineering Center at Ohio State University [65] and engineers without borders [66, 67] provided data from a variety of developing community programs. The real-world tangible nature of SL design projects is a significant motivator, in addition to making a positive difference.

6. Assessing community impacts of SL design projects

Service-learning has co-equal goals of benefits to community partners and student learning. Assessment is needed to demonstrate whether SL design projects have met these goals. SL projects may have impacts at the individual, organizational, community, or system scales [68]. Jiusto and Vaz [68] present a model that considers these impacts to both communities and academics, which can inspire instructors considering the use of SL as a design pedagogy to think beyond immediate impacts. This broader systems-level perspective can include potential project outcomes such as improvements in the health and well-being of community partners, while recognizing how these outcomes might contribute to enhancing community sustainability or social cohesion. Identifying impacts of interest in partnership with all stakeholders is the first step in developing a plan to assess these impacts.

In practice, SL has often focused its assessment efforts on student learning and less on evaluating the impacts on community partners and communities; this imbalance is evident both for SL in the context of engineering design and SL more broadly [69–71]. Reynolds [72] provides a critical review of literature on community perspectives on service-learning, and conducted research on the perspectives of the international partner community in Nicaragua on their partnership with the College of Engineering at Villanova University. Although this was a research project, assessment lessons can be learned. Observations, interviews with community organization representatives, interviews with community residents, and document reviews were conducted. Community partners confirmed the tangible results of improved access to clean water and healthcare which saved lives, but also described trust, a sense of pride, and connections/awareness as important outcomes. The community also had less positive perceptions that included feeling like their community was a laboratory for students. The community also had goals toward student learning, including shifting students' perspectives from helping to learning and having a responsibility to others.

These findings represent the particular ways in which SL projects were conducted in this instance and their specific community partners, and should not be generalized. However, these important insights provide an example of the types of outcomes that assessment can illuminate. Others have also used interviews [73, 74] and surveys [14, 74, 75] to assess community partner satisfaction and other perspectives on SL engagement. Readers are encouraged to consult participatory action research models [76] to learn more about the process of planning, executing, and evaluating projects together; communication, transparency, and shared power in decision-making are hallmarks of these approaches.

Design projects and their products should be monitored over time to evaluate sustainability and long-term impacts. This process is easier for projects in local communities and more challenging for international projects, but is critical in all cases. SL projects could model practices and processes used in international development work for monitoring and evaluation (M&E), which typically include mixed-methods [77]. The community and/or students can be involved in monitoring the designed systems, and can work together to resolve any issues that are identified. On-going collaboration with groups charged with monitoring and evaluation is also a strategy. For example, with the LSU Community Playground Project [33, 41], once community-designed playgrounds are built at public schools, a company that subcontracts with the school system to provide grounds and maintenance services to the schools takes over the maintenance of the playgrounds. On-going communication among the playground project, the school system, and the company ensures that playgrounds are re-designed, built, and maintained based on need.

7. Conclusions

Done well, service-learning based engineering design can yield a rich array of benefits for engineering students and communities. However, faculty must carefully plan their course and partnership in order to achieve the full potential of SL-based design. Engineering faculty and students should enter into the design process from a mindset of humility and listening, being respectful, and embracing the expertise of the community. This positioning is often different from the techno-centric, "expert" perspective that pervades engineering. To instill this human-centered or empathic design perspective in students, their first formal education on the engineering design process should promote these views. This approach can perhaps grow into participatory design in the senior year. One challenge is the fact that many engineering faculty members have not previously experienced such approaches, either during their education and training, or in practice. Fortunately, the literature provides rich examples for faculty to draw from to implement this methodology in their own courses.
We believe that best practices in service-learning in engineering design make our students better engineers and enables our profession to fulfill its highest purposes.

Acknowledgements

The authors gratefully acknowledge our community partners and academic collaborators over the years, from whom we have learned and grown professionally and personally.

This work was partly supported by the USDA National Institute of Food and Agriculture (LAES project #94261). Publication of this chapter was funded by the University of Colorado Boulder Libraries Open Access Fund.

Conflict of interest

The authors declare that they have no conflict of interest related to this work.

Author details

Angela R. Bielefeldt[1]* and Marybeth Lima[2]

1 University of Colorado Boulder, Boulder, CO, USA

2 Louisiana State University, Baton Rouge, LA, USA

*Address all correspondence to: angela.bielefeldt@colorado.edu

References

[1] Amadei BA. Global engineering for a small planet. In: Keynote Presentation at the Zone IV Conference of the American Society for Engineering Education (ASEE); 25-27 March 2018; Boulder, CO

[2] Conley SN, Foley RW, Gorman ME, Denham J, Coleman K. Acquisition of T-shaped expertise: An exploratory study. Social Epistemology. 2017;**31**(2):165-183

[3] Rogers P, Freuler RJ. The T-shaped engineer. In: Proceedings of the American Society for Engineering Education (ASEE) Annual Conference & Exposition; 14-17 June 2015; Seattle, WA; Washington, DC: ASEE 2015. 18 p

[4] Swan C, Paterson K, Bielefeldt A. Community engagement in engineering education as a way to increase inclusiveness. In: Johri A, Olds B, editors. Cambridge Handbook of Engineering Education Research (CHEER). New York: Cambridge University Press; 2014. pp. 357-372

[5] Bringle R, Hatcher J. A service-learning curriculum for faculty. Michigan Journal of Community Service Learning. 1995;**2**:112-122

[6] Boyer E. Scholarship Reconsidered: Priorities of the Professoriate. Princeton: Carnegie Foundation for the Advancement of Teaching; 1990

[7] Boyer E. The scholarship of engagement. Journal of Public Service and Outreach. 1996;**1**(1):11-20

[8] Harkavy I. Service-learning and the development of democratic universities, democratic schools, and democratic good societies in the 21st century. In: Welch M, Billig S, editors. New Perspectives on Service-Learning: Research to Advance the Field. Greenwich: Information Age Publishing; 2004. pp. 3-22

[9] Eyler J, Giles D. Where's the Learning in Service-Learning? San Francisco: Jossey-Bass; 1999. ISBN: 0470907460

[10] Zlotkowski E, editor. Service-Learning in the Disciplines. Sterling: Stylus Publishing; 1998. 20+ Volume series on service-learning in various professions; entire series. Available from: http://styluspub.com/Books/SearchResults.aspx?str=service-learning+in+the+disciplines

[11] Kuh GD. High-Impact Educational Practices: What they Are, Who Has Access to them, and why they Matter. Washington, DC: American Association of Colleges & Universities; 2008. 34 p

[12] Heffernan K. Fundamentals of Service-Learning Course Construction. Providence, RI: Campus Compact, Brown University; 2001

[13] Coyle E, Jamieson L, Sommers L. EPICS: A model for integrating service-learning into the engineering curriculum. Michigan Journal of Community Service Learning. 1997;**4**:81-89

[14] Coyle EJ, Jamieson LH, Oakes WC. EPICS: Engineering projects in community service. International Journal of Engineering Education. 2005;**21**(1):139-150

[15] EPICS—Engineering Projects in Community Service. EPICS Consortium Institutions [Internet]. 2018. Available from: https://engineering.purdue.edu/EPICS/university/institutions [Accessed: Nov 21, 2018]

[16] Tsang E, editor. Projects that Matter: Concepts and Models for Service-Learning in Engineering. Washington, DC: American Association for Higher Education; 2000

[17] Swan CW, Duffy JJ, Paterson K, Bielefeldt AR, Pierrakos O. The

EFELTS project—Engineering faculty engagement in learning through service. In: Proceedings of the American Society for Engineering Education (ASEE) Annual Conference & Exposition; 26-29 June 2011; Vancouver. Washington, DC: ASEE. 14 p

[18] McCahan S, Ault HK, Tsang E, Henderson MR, Magleby SP, Soisson A. A multi-dimensional model for the representation of learning through service activities in engineering. In: Proceedings of the American Society for Engineering Education (ASEE) Annual Conference & Exposition; 26-29 June 2011; Vancouver. Washington, DC: ASEE. 16 p

[19] Colledge T, editor. Special issue: Opportunities and barriers to integrating service learning into engineering education. In: International Journal for Service Learning in Engineering. 2014. 169 p. Available from: https://ojs.library.queensu.ca/index.php/ijsle/issue/view/461 [Accessed: Nov 21, 2018]

[20] Colledge T, editor. Special issue: University engineering programs that impact communities: Critical analyses and reflection. In: International Journal for Service Learning in Engineering. 2015. 562 p. Available from: https://ojs.library.queensu.ca/index.php/ijsle/issue/view/522 [Accessed: Nov 21, 2018]

[21] Zoltowski CB, Oakes WC, Cardella ME. Students' ways of experiencing human-centered design. Journal of Engineering Education. 2012;**101**(1):28-59

[22] IDEO. The Field Guide to Human-Centered Design. 1st ed. Canada: IDEO.org; 2015

[23] Crawley EF, Malmqvist J, Ostlund S, Bordeur DR, Edstrom K. Rethinking Engineering Education: The CDIO Approach. 2nd ed. Cham: Springer; 2014

[24] Lucena J, Schneider J, Leydens J. Engineering and sustainable community development: Critical pedagogy in education for "engineering to help". In: Proceedings of the American Society for Engineering Education (ASEE) Annual Conference & Exposition; 20-23 June 2010; Louisville, KY. Washington, DC: ASEE. 16 p

[25] Leydens JA, Lucena JC. Engineering Justice: Transforming Engineering Education and Practice. Hoboken, New Jersey: Wiley-IEEE Press; 2017. 304 p. ISBN: 978-1-118-75730-7

[26] Lambrinidou Y, Rhoads WJ, Roy S, Heaney E, Ratajczak GA. Ethonography in engineering ethics education: A pedagogy for transformational listening. In: Proceedings of the American Society for Engineering Education (ASEE) Annual Conference & Exposition; 15-18 June 2014; Indianapolis, IN. Washington, DC: ASEE. 27 p

[27] Hess JL, Fila ND. The manifestation of empathy within design: Findings from a service-learning course. CoDesign. 2016;**12**(1-2):93-111

[28] Catalano GD. Engineering education: Moving toward a contemplative service paradigm. In: Proceedings of the American Society for Engineering Education (ASEE) Annual Conference & Exposition; 26-29 June 2016; New Orleans, LA. Washington, DC: ASEE. 11 p

[29] Dean JH, Van Bossuyt DL. Breaking the tyranny of the semester: A phase-gate sprint approach to teaching Colorado School of Mines students important engineering concepts, delivering useful solutions to communities, and working on long time scale projects. International Journal for Service Learning in Engineering. 2014;**9**:222-239. https://ojs.library.queensu.ca/index.php/ijsle/issue/archive

[30] Paterson KG, Bielefeldt AR, Swan CW, Rulifson G, Kazmer D, Pierrakos O. Designing value into engineering learning through service activities using a blueprint model. International Journal for Service Learning in Engineering. 2013;**8**:64-83

[31] Nelson LA. Scaffolding undergraduate engineering design education with the wellbeing framework. In: Proceedings of the American Society for Engineering Education (ASEE) Annual Conference & Exposition; 10-13 June 2012; San Antonio, TX. Washington, DC: ASEE. 11 p

[32] Traum MJ, Howell DA, Newman LC. Engineering design, project management, and community service connected through servant leadership. In: Proceedings of the American Society for Engineering Education (ASEE) Annual Conference & Exposition; 23-26 June 2013; Atlanta, GA. Washington, DC: ASEE. 13 p

[33] Lima M. Building Playgrounds, Engaging Communities: Creating Safe and Happy Places for Children. Baton Rouge: LSU Press; 2013

[34] Stoecker R. Liberating Service Learning and the Rest of Higher Education Civic Engagement. Philadelphia: Temple University Press; 2016

[35] Community-Campus Partnerships for Health. Achieving the Promise of Authentic Community-Higher Education Partnerships: Community Partners Speak Out; 24-26 April 2006. Racine, WI: Wingspread Conference Center. Seattle: CCPH; 2007. Report available from: www.ccph.info

[36] Portland State University. A guide to reciprocal community-campus partnerships. In: Proceedings from Portland State University's Partnership Forum. March 6-8, 2008. 8 p. Available from: https://depts.washington.edu/ccph/pdf_files/Guide_corrected_041808.pdf [Accessed: Nov 22, 2018]

[37] Mitchell T. Traditional versus critical service-learning: Engaging the literature to differentiate two models. Michigan Journal of Community Service Learning. 2008;**14**(2):50-65

[38] Schön D. The Reflective Practitioner: How Professionals Think in Action. New York: Basic Books; 1983

[39] Lima M, Oakes WC. Service-Learning: Engineering in your Community. 2nd ed. New York: Oxford University Press; 2014

[40] Bielefeldt AR, Paterson KG, Swan CW. Measuring the value added from service learning in project-based engineering education. International Journal of Engineering Education. 2010;**26**(3):535-546

[41] Lima M. The LSU community playground project: Reflections on 16 years of an engineering service-learning program. International Journal for Service-Learning in Engineering. Special Issue: University Engineering Programs that Impact Communities: Critical Analyses and Reflection. 2014:492-508

[42] Arrazattee C, Lima M, Lundy L. Do university communications about campus-community partnerships reflect core engagement principles? Michigan Journal of Community Service Learning. 2013;**20**(1):41-52

[43] Celio CI, Durlak J, Dymnicki A. A meta-analysis of the impact of service-learning on students. The Journal of Experimental Education. 2011;**34**(3):164-181

[44] International Engineering Alliance (IEA). Graduate Attributes and Professional Competencies. Version 3. 21 2013. 16 p

[45] ABET. Engineering Accreditation Commission. Criteria for Accrediting Engineering Programs. Effective for Reviews during the 2018-2019 Accreditation Cycle. Baltimore, MD: ABET; 2017

[46] Bielefeldt AR, Polmear M, Knight D, Swan C, Canney N. Incorporation of ethics and societal impact issues into senior capstone design courses: Results of a national survey. In: Proceedings of the American Society for Engineering Education (ASEE) Annual Conference and Exposition; 25-28 June 2017; Columbus, OH. Washington, DC: ASEE. 19 p

[47] Titus C, Zoltowski CB, Huyck M, Oakes WC. The creation of tools for assessing ethical awareness in diverse multi-disciplinary programs. In: Proceedings of the American Society for Engineering Education (ASEE) Annual Conference and Exposition; 26-29 June 2011; Vancouver. Washington, DC: ASEE. 17 p

[48] National Academy of Engineering. EPICS (Engineering Projects in Community Service) program. In: Infusing Real World Experiences into Engineering Education. Washington, DC: National Academies Press; 2012. 35 p

[49] American Society of Civil Engineers (ASCE). Civil engineering body of knowledge. In: Preparing the Future Civil Engineer. 3rd ed. Reston, VA: ASCE, Spring; 2019. (in press). https://www.asce.org/civil_engineering_body_of_knowledge/

[50] Stanlick S, Sell M. Beyond superheroes and sidekicks: Empowerment, efficacy, and education in community partnerships. Michigan Journal of Community Service Learning. 2016;**23**(1):80-84

[51] Wang L, Carroll T, Delaine D. A pilot study of the development of empathy within a service-learning trip from a qualitative perspective. In: Proceedings of the American Society for Engineering Education (ASEE) Annual Conference & Exposition; 24-27, 2018; Salt Lake City, UT. Washington, DC: ASEE. 22 p

[52] Charyton C, Jagacinski RJ, Merrill JA, Clifton W, DeDios S. Assessing creativity specific to engineering with the revised creative engineering design assessment. Journal of Engineering Education. 2011;**100**(4):778-799

[53] Shelby R, Ansari F, Patten E, Pruitt L, Walker G, Wang J. Implementation of leadership and service learning in a first-year engineering course enhances professional skills. International Journal of Engineering Education. 2013;**29**(1):1-14

[54] Ariely S, Banzaert A, Wallace D. Mechanisms for implementing service learning: Analysis of efforts in a senior product design class in mechanical engineering. In: Proceedings of the American Society for Engineering Education (ASEE) Annual Conference & Exposition; 12-15 June 2005; Portland, OR. Washington, DC: ASEE. 11 p

[55] Christy AD, Lima M. Developing creativity and multidisciplinary approaches in teaching engineering problem-solving. International Journal of Engineering Education. 2007;**23**(4):636-644

[56] Davis I, Plumblee JM, Brown A, Vaughn D. Encouraging innovation through design in resource constrained environments. In: Proceedings of the American Society for Engineering Education (ASEE) Southeastern Section Conference.... Daytona Beach FL. March 2018. p. 4-6. http://www.asee-se.org/proceedings/ASEE2018/papers2018/50.pdf

[57] Huff JL, Abraham DM, Zoltowski CB, Oakes WC. Adapting curricular models for local service-learning to international ommunities. In:

Proceedings of the American Society for Engineering Education (ASEE) Annual Conference & Exposition; 10-13 June 2012; San Antonio, TX. Washington, DC: ASEE. 14 p

[58] Huff JL, Zoltowski CB, Oakes WC. Preparing engineers for the workplace through service learning: Perceptions of EPICS alumni. Journal of Engineering Education. 2016;**105**(1):43-69

[59] Knizley AA, Coleman TA. Implementing service-learning into an introductory mechanical engineering course. In: Proceedings of the American Society for Engineering Education (ASEE) Southeast Section Conference. 2014. 6 p

[60] Mehta K, Zappe S, Brannon ML, Zhao Y. An educational and entrepreneurial ecosystem to actualize technology-based social ventures. Advances in Engineering Education. 2016;**5**(1):39

[61] Pinnell M, Daniels M, Hallinan K, Berkemeier G. Leveraging students' passion and creativity: ETHOS at the University of Dayton. International Journal for Service Learning in Engineering. 2014;**9**:180-190

[62] Duffy JJ, Barrington L, Munoz MZH. Attitudes of engineering students from underrepresented groups toward service-learning. In: Proceedings of the American Society for Engineering Education (ASEE) Annual Conference & Exposition; 26-29 June 2011; Vancouver, BC. Washington, DC: ASEE. 29 p

[63] Matusovich HM, Oakes W, Zoltowski CB. Why women choose service-learning: Seeking and finding engineering-related experiences. International Journal of Engineering Education. 2013;**29**(2):388-402

[64] Dzombak R, Mouakkad S, Mehta K. Motivations of women participating in a technology-based social entrepreneurship program. Advances in Engineering Education. 2016;**5**(1):29

[65] Bixler G, Campbell J, Dzwonczyk R, Greene HL, Merrill J, Passino KM. Humanitarian engineering at the Ohio State University: Lessons learned in enriching education while helping people. International Journal for Service Learning in Engineering. 2014;(Fall, special edition):78-96

[66] Litchfield K, Javernick-Will A. Investigating gains from EWB-USA involvement. Journal of Professional Issues in Engineering Education and Practice. 2014;**140**:0413008-1-0413008-9

[67] Bielefeldt AR. Attracting women to engineering that serves developing communities. In: Proceedings of the American Society for Engineering Education (ASEE) Annual Conference & Exposition; 18-21 June 2006; Chicago, IL. Washington, DC: ASEE. 10 p

[68] Jiusto S, Vaz RF. Understanding impacts: Community engagement programs and their implications for communities, campuses and societies. In: Filho WL, Nesbit S, editors. New Developments in Engineering Education for Sustainable Development. Switzerland: Springer; 2016. pp. 125-138

[69] Bielefeldt AR, Paterson K, Swan C, Duffy JJ, Pierrakos O, Canney NE. Engineering faculty engagement in learning through service summit: Best practices and affinity mapping. In: Proceedings of the American Society for Engineering Education (ASEE) Annual Conference & Exposition; 10-13 June 2012; San Antonio, TX. Washington DC: ASEE. 16 p

[70] Hartman E, Kiely R, Smedley CT, Reynolds N, Mather MC, Gregory M. Learning from community: Community outcome assessment best practices and

insights in global service-learning. In: International Association for Research on Service-Learning and Community Engagement Annual Conference; 6-8 Nov 2013; Omaha NE. Cited in: Campus Compact. Available from: https://compact.org/learning-community-community-outcome-assessment-best-practices-insights-global-service-learning/. [Accessed: Nov 28, 2018]

[71] Blouin DD, Perry EM. Whom does service learning really serve? Community based organizations' perspectives on service learning. Teaching Sociology. 2009;**37**(April):120-135

[72] Reynolds NP. Is international service-learning win-win?: A case study of an engineering partnership [PhD dissertation]. Ann Arbor MI: Temple University; 2016 ProQuest Number 10112434. 181 p

[73] Duffy J, Barrington L, West C, Heredia M, Barry C. Service-learning integrated throughout a college of engineering (SLICE). Advances in Engineering Education. 2011;**2**(4):23

[74] Zoltowski CB, Oakes WC. Learning by doing: Reflections of the EPICS program. International Journal for Service Learning in Engineering. 2014;**9**:1-32

[75] Cowan D. A client-based assessment tool for service learning projects. In: Proceedings of the American Society for Engineering Education (ASEE) Annual Conference & Exposition; 22-25 June 2008; Pittsburgh, PA. Washington, DC: ASEE. 16 p

[76] Pain R, Whitman G, Milledge D, Lune Rivers Trust. Participatory Action Research Toolkit: An Introduction to Using PAR as an Approach to Learning, Research, and Action. Available from: http://communitylearningpartnership.org/wp-content/uploads/2017/01/PARtoolkit.pdf [Accessed: Nov 30, 2018]

[77] Bamberger M, Rao V, Woolcock M. Using Mixed Methods in Monitoring and Evaluation: Experiences from International Development. The World Bank; 2010 Policy Research Working Paper 5245. 30 p

Chapter 3

Divergent Abduction Model and Its Convergent Interaction in Knowledge Production

Manuel Vicente Castillo Guilarte

Abstract

To date, the abductive is maintained as a method by which premises or isolated conjectures are generated from deductive and inductive methods, whose basis is convergence and divergence, respectively. The vast majority of authors do not relate the abductive method to the scientific method. Starting from a descriptive methodology and deductive analysis, the objective of the work is to propose a new pattern of extended use of the abductive method and its interaction with the inductive method based on divergence and, in this way, schematically and organically achieve a convergent procedure of support for the production of new knowledge. In this chapter, a model is established integrating the abductive n times with the divergence, and, logically, unifying and integrating are reduced until reaching a solution by convergence and, in this way, to achieve alternative processes of preparation and academic training to support and comply with the requirements of promotion and doctoral thesis in relation to generating new knowledge, as established in the rules that govern it in Venezuela.

Keywords: abductive divergence, abduction, convergence interaction

1. Introduction

The production of knowledge has always been a necessity, and it is the final fruit of research and education. The search for knowledge has a very clear objective and concludes in the contribution that occurs when there is an invention or an innovation, but, when analyzing the different countries, the difference between being and not developing depends on how their level is in science and technology. Venezuela as a country has three major problems: the low production of knowledge, the syndrome of anything but thesis in doctoral studies, and the elimination of the requirements of having a doctorate for promotion to the last categories in some universities. Starting from this problematic, the present chapter is developed in which, centered on a deductive documentary methodology, a conceptual framework of abduction and divergence is defined, which are integrated to establish a convergent model of knowledge production with its corresponding stages. In Venezuela, there is a legislation published in the Official Gazette [1] that states in article 26: "Doctoral studies are aimed at training for the realization of an original work that constitutes a significant contribution to the knowledge of a specific area of knowledge." To illustrate, in Venezuela, historically, the doctorate consists of constructing an original contribution relevant to science, for which the generation of knowledge is mandatory in the doctorate; the rest of the Venezuelan legislation speaks and defines research but without involving it with patents, and, although there are universities that are generally responsible for educating and training researchers, only the doctoral thesis at the graduate level is developed with the aim of generating new knowledge. Additionally, in Venezuela, there is a National Universities Law published in Official Gazette N° 1.429 of September 8, 1970, in article 87 which establishes that they are ordinary members of teaching and research staff, instructors and assistant, aggregates, associates, and full professor, known as the ranking or location of teachers according to the merits and time spent on average of 4 years in each category. To ascend from one level to another in the universities is defined in article N° 89 which says: "The Ordinary members of the teaching and research staff will be located and will rise in the ranks according to their credentials or scientific merits and their years of services. To ascend from one category to another in the ranking, it will be necessary, in addition, to present for the consideration of a jury appointed for that purpose an original work as a credential of merit." The important thing of this article is that the work of promotion to ascend to the different levels of the ladder must be original; but as time went by and after 24 years, a modification to the Law of Universities published in Official Gazette No. 35, 708 of May 11, 1995, called Standards on the Teaching and Research Staff Scale of National Universities, and in article N° 5, it establishes: "The work of promotion required by the Law of Universities must be original, novel and untold published during the time in which the teacher remained in the previous ladder to which one aspires"; this article introduces and raises the requirement that, besides being original, must comply with being novel, which leads to new verifiable knowledge worldwide, and the unpublished is that it has not been published previously; the legislators seek to establish the obligatory nature of the ordinary teachers of the universities, to produce new knowledge on the frontier of science, which in practice is invention and innovation.

After an analysis of the problematic, the most important of the writing is the divergent adductive model of n hypotheses, all with the same probability, where infinite conjectures of a fact or event are defined, and then from a convergent step-by-step process, look for the possible solution, and in the end, a general model that integrates both approaches in a single scheme is presented.

2. Cause of Low Production of Knowledge

There is a great variety of problems related to the low production of knowledge in Venezuela, the most outstanding being:

- The percentage of students who graduate from doctorates as a source of knowledge production is approximately 10% compared to the income, which defines that there is a low efficiency in this schooling in Venezuela; this oscillates between 0 and 15% per year, with a unique atypical value of a University with a 95% ratio of graduates of the total income. Everything is based on the information provided by the universities to the National Graduate Advisory Council [2]; the cause is known as everything except thesis [3], which means finishing school and doctoral thesis project, but it does not end with the delivery and public defense of the said thesis.

- The actors that are required to produce knowledge are university teachers, and they are not fulfilling their task. By law they are obliged to investigate and, also, to climb the ladder in the categories of associate and holder; they must have a doctorate, but, in the great majority of the universities of the country, an exceptionality persists so as not to fulfill the requirement of the title of doctor [4], which increases inefficiency and allows people to reach the highest levels of teaching without academic credentials. The experts in generating patents par excellence are the university professors, but if these do not fulfill their obligations, they do not possess the experience to support others, which turns into a vicious circle.

- It does not exist an orientation, models, or patterns adapted to our reality that supports the generation of knowledge, being tutors who should have a high level of expertise and experience in the subject of doctoral thesis, publication, incubation, prototypes, and patents and their inclusion in the application to solve problems. The different methodologies known worldwide are not giving real benefits in Venezuela, and both the level of scarcity of financial resources for research projects and low salaries become factors of low motivation; all the above is reflected in the world records which places Venezuela among the group with the lowest number of patents per year.

3. Objective

Define a basic model of abduction and divergence interaction that allows systemically to model with the variables involved in a convergent sequential pattern to increase knowledge production in Venezuela.

4. Methodology

The research is based on a study and analysis of the information developed in other areas of knowledge and its application in the subject, to relate them to the dynamics over time of the problem and the generation of knowledge; starting from the use of a descriptive methodology, of deductive analysis with emulation of different sciences, the convergence and divergence engineering is established as a tool to model the scope of knowledge generation. The research begins by locating and studying the sources of information to subsequently analyze and interpret in a critical and separate way the ideas surrounding the topic, to relate them as variables that intervene in a basic model of knowledge generation.

5. Theoretical support

Abduction is a term from the Latin abductio and is composed of the words abs, from afar, and ducere, to carry. The movement by which a member or an organ moves away from the median plane divides the body in two symmetrical parts. "The abductor is the muscle that serves for abduction" is a type of reasoning that from the description of a fact or phenomenon offers or arrives at a hypothesis, which explains the possible reasons or reasons for the fact through the premises obtained. Charles Sanders Peirce calls it a conjecture [5]. That conjecture seeks to be, at first sight, the best explanation or the most probable one.

Abductive thinking or reasoning is a type of reasoning that from a fact or phenomenon is reached a hypothesis, which explains the possible reasons or reasons for establishing a premise called conjecture. That conjecture seeks to be, at first sight, the best explanation or the most probable one. Aristotle investigated seductive reasoning in his First Analytics. According to Aristotle, abductive reasonings are syllogisms where the premises only give a certain degree of probability to the conclusion. According to Peirce, abduction is something more than a syllogism: it is one of the three forms of reasoning together with deduction and induction. Charles S. Peirce (1903) [5] defines abduction: "Abduction is the process by which an explanatory hypothesis is formed, it is the only logical operation that introduces a new idea." Abduction can be understood as a form of logical inference. In abduction to understand a phenomenon, a rule that operates in the form of hypotheses is introduced to consider the possible result as a particular case within that rule; in other words, in the case of a deduction, a conclusion is obtained of a Premise "p," while abductive reasoning consists of explaining "c" by means of "p" considering p as explanatory hypothesis [6]. Abduction is characterized, then, as a creative process, as it generates the new ideas, while the deduction derives knowledge from the one that has already been validated previously, and the induction, for its part, is limited to checking it in. Abduction, allows the identification of indications to which something corresponds and the reasons for their appearance, from which a series of consequences can be extracted.

Divergence is a word that comes from Latin "divergens" which means divergent. It means action and effect of diverging, progressive separation of two or more lines or surfaces; diversity of disagreements, diversity of opinions, as an example we have in mathematics the divergence of a vector field on a surface and the infinite series that does not converge [7].

Divergent thinking: Throughout history there are many treaties that established the classification of productive thinking in two groups: divergent and convergent. Divergent thinking is pointed by the authors as a generator of different solutions to a problem, in different directions, seeking the best; confronting the new and with a tendency to creativity and ingenuity, it has also been called lateral thought, by De Bono in 1970, as well as holistic thinking defined by Jan Smuts in 1927, in which the analysis is of the whole and not of its parts, including as a term "see the forest and not the tree." There is a mechanism associated with creativity, which is retrospective perception and intuition; according to the story, it is established that the great discoveries have been produced by changes of schemes through casual observations, accidents, errors, and humor. Rio Pérez, in 2002, characterized the creative thinking of an individual and related it to the intellectual structure of the subject, their characteristics, and their mental capacities [8]. At the beginning, human development focused on intelligence, knowledge, memory, and logical or convergent thinking, and, later, divergent production was incorporated [9]. The production of knowledge, in the first instance, is the interaction of vertical and lateral thought closely linked, where lateral thinking is the one that increases the efficiency of vertical thinking.

6. Case study

To define the case study, it is formulated in the Seminar on Management in Science and Technology, of the Doctorate of Management of Research and Development, during the academic periods of the second of 2011 to the first of 2014, with an average of seven students, belonging to the Area of Management of Research and Development, of the Commission of Postgraduate Studies of the UCV, to which a question was posed to the trainees: Is there a ticket on the floor? Unknown, what do you think about it? Do you formulate all the hypotheses related to the case? **Table 1** shows the relation of the number of students and the total of hypotheses that they managed to define for 5 min, which is an application of establishing premises when an event occurs, with a total of 40 curative participants during six semesters.

People	Hypothesis
1	1
18	2
15	3
2	4
3	5
1	6

Table 1.
Relationship and participants responses.

The challenge was to produce the greatest number of hypotheses; in many cases, the abductions are but the spontaneous conjectures of reason. In order for these hypotheses to emerge, the imagination and instinct must compete. Abduction is like a flash of understanding, a jump over the known; for abduction, it is necessary to leave the mind free. Based on the same events, then the students are illustrated with a variety of hypotheses of equal probability; out of a total of 100, there are at least 35 equally likely assumptions and they are:

- It's false?
- Is the evil?
- Will it be paving?
- Is it terror?
- What brought a bird?
- Will it be a student?
- Will it be a professor?
- Will it be a pump?
- Will it be a joke, with a hidden camera?
- Will they be filming?
- Will you have owner?
- Is it good luck?
- Will it be embraced?
- Will it have poison?
- Will it have a bar code?
- Will it have GPS?
- Will you have intelligent chip?

- Will it be an alarm activator?
- Will you be in a research?
- Will it be a hologram?
- Will it be mine?
- Is it a projection?
- Will it be contaminated from drug?
- Is it a transparent touch screen?
- Is it an optical illusion?
- Will it be printed on the floor?
- Is it glass?
- Will it be plastic?
- Is it a photo?
- Is it a joke?
- Is it a proof of ethics?
- Is it a proof of honesty?
- Is it a proof of success?
- Is it a proof of education?

7. Model of generation of knowledge

Starting from the definition of abduction and divergence [10], the model will be presented in five parts to increase understanding; each phase will have a comment and the basis of the construction. The first phase consists of applying abduction infinitely starting from a central axis and according to the abduction capacities; a scenario will be produced in which all the hypotheses are distributed, and it is illustrated in **Figure 1**. In a broad way, it is possible to name this new concept as "Conductive Abductive Model of Infinite Conjectures and Hypotheses" considering itself as a novel contribution that aims to broaden the scope of the traditional abductive method defined by Pierce.

The second step, once infinite hypotheses are defined, is to order all of them according to some preestablished macrovariables, which will depend on the experience and training of each person; however, part of these may be:

- Morals, ethics.
- Cultural, habits, customs.

- Formation, education instruction.
- Family, society, country.
- Science, technology,
- Invention, innovation.
- Sentiment, intuition, religion, creed.
- Development, underdevelopment, poverty, wealth.
- Link, irreality, fiction, superstition, reality.
- Heat, threat, insecurity.
- Equipment electrical, mechanical, communications.
- Defense, protection, integrity.
- Politics, government, doctrines.
- Experience, techniques, tools, skill, etc.

The hypotheses are grouped according to each variable, which is illustrated in **Figure 2**.

In the third phase of the model, we seek to establish the different relationships between the hypotheses and their relationship with the variables to which they are associated, which is presented in **Figure 3**, and, in this way, a new group of integrated hypotheses is established and selected according to the different preestablished criteria, which gives a new higher level, in which at least 30% less participate, reducing the total number of hypotheses.

The fourth phase of the model is presented, which combines and establishes a process of integration, simplification, and reduction until it reaches at least one, two, or three possible solutions with characteristics; theoretical logics that are novel are presented in **Figure 4**. In this figure, the convergent model is characterized, from the general to the particular, from the very broad to the simple.

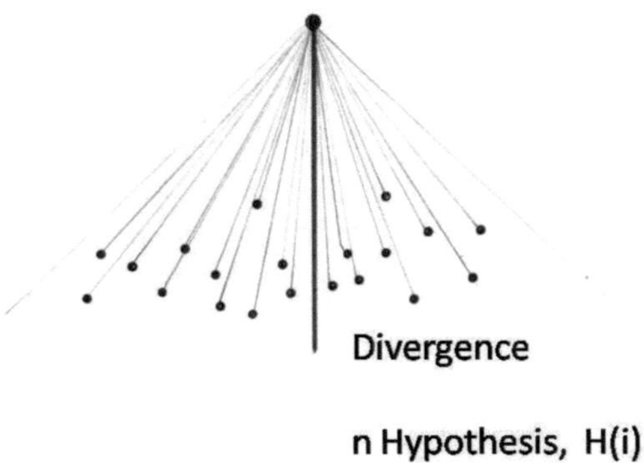

Figure 1.
Infinite application of abduction.

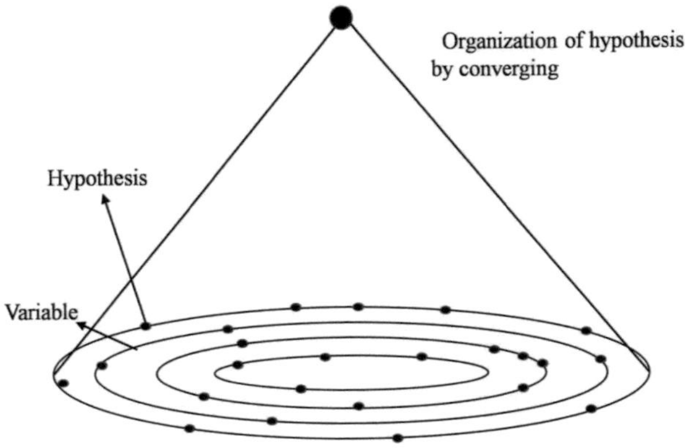

Figure 2.
Association of hypotheses in variables.

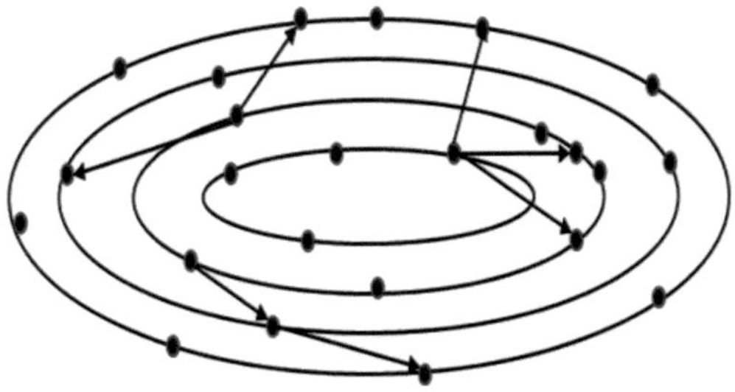

Figure 3.
Relationships and links of the hypothesis with others and the new preset variables.

The model has been illustrated in three initial figures, to explain the divergence of n abductions and the convergence of these in a certain pattern or technique; harmonizing the four previous figures integrated into a final model is presented in **Figure 5**, and it is called Integrated Model of Generation of Knowledge, which is characterized by the following:

- Everything begins with a problem and ends with one or several novel solutions that depend on the cognitive capacity of each actor.
- From a problem, infinite hypotheses are generated producing a three-dimensional divergence, and the size, range, spectrum, and novelty will be proportional to the total skills of the individual involved, and the farther they can go in different directions, the closeness to produce new knowledge it will be greater.
- Having as origin, divergent model in two or three dimensions the only way to reach a defined point as a solution to the problem is to apply a convergence, which requires ingenuity and engineering; as an organized and systematic procedure, it will be specific to each actor if based on their experiences, schooling, logic, and reasoning.

In the field of engineering, specifically the area of networks, a simile could be proposed, the distribution and water collection system, in the understanding that we start from a drinking water dam; as a starting point, we will make a distribution system in such a way that we start with pipes of high diameter to reach the smallest as a sink and start at a point and takes the water to infinite points, achieving and

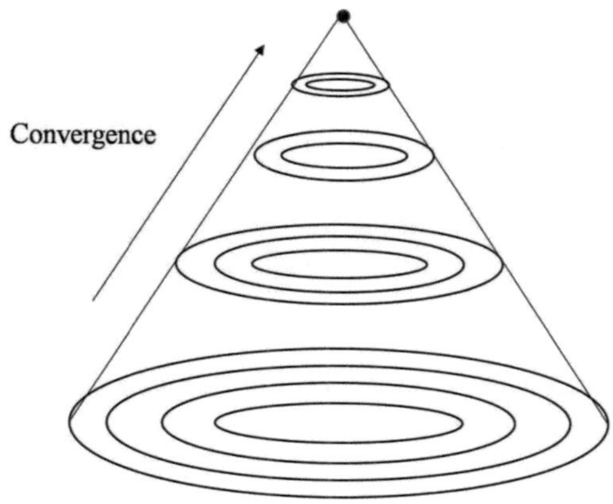

Figure 4.
General model of knowledge production.

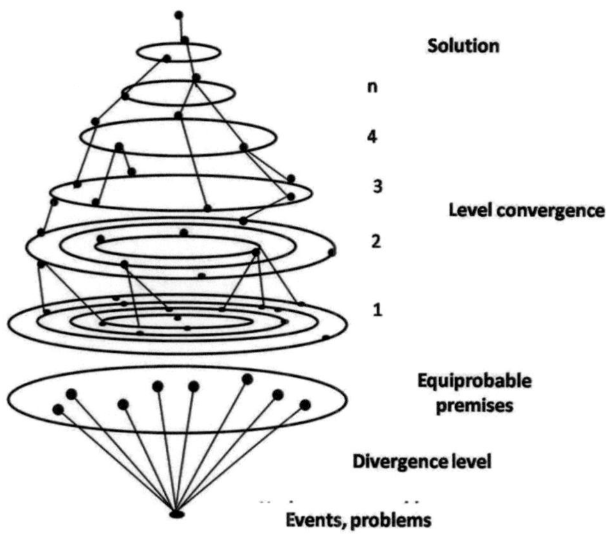

Figure 5.
Integrated generation of knowledge model.

applying a divergent system; once the water is used, it is transformed into waste or wastewater, which must be channeled and taken to a sewage treatment plant and then incorporated into the dam or river, starting from infinite points and reaching one, which is a convergent system. This comparison is a way of understanding how the model of generation of raised knowledge works, part of a point called fact or event, and of producing infinite hypothesis, and then convergence starts, simplifying until arriving at a point as final selection of hypothesis.

8. Conclusions

- In this work, we present a novel contribution that aims to redefine the traditional abductive method, called "Conductive Abductive Model of Infinite Conjectures and Hypotheses."
- The proposed model is a new approach that integrates the scientific method with the modified abductive method, establishing a new sequential form to search for and obtain alternatives in the generation of new theories by convergences.

- The model of divergent abduction linked to the convergent process of selection is transformed into a procedure to generate and produce new knowledge, which contributes to improvement for graduate students, which, using the model, guides them to success and search for new theories.
- The model systematized as part of a seminar, in the schooling of doctoral studies, can be the basis for training the students and strengthening their skills for knowledge production.
- The model achieved is novel, but its deduction was obtained with the same principles set out in this work.

Author details

Manuel Vicente Castillo Guilarte
Department of System Engineering, National Experimental Polytechnic University UNEXPO, Caracas, Venezuela

*Address all correspondence to: mvcg22@gmail.com

References

[1] General Regulations for Postgraduate Studies, Official Gazette of the Bolivarian Republic of Venezuela No. 37628; 2001

[2] CCNPG, Postgraduate Directory of Venezuela, National Council of Universities. CNU-MPPEU, Caracas, Venezuela; 2011. Available from: http://www.ccnpg.gob.ve/directorio_nacional/busqueda_ava.asp?buscar=no&orden1=1&orden2=4

[3] Humberto E. How Do You Write to Thesis. Barcelona: Editorial Gedisa; 2011

[4] Castillo M. Prospects of teaching profession in universities and its relationship with the engineering in Venezuela. In: Eighth LACCEI Latin American and Caribbean Conference for Engineering and Technology; Arequipa Peru; 2010. Paper 163

[5] De Bono E. Lateral Thinking a Text Book of Creativity. UK: Mica Management; 1970

[6] Del Río Pérez J. Review of psychological research on creativity. Observatory (OBS). 2007;**1**(2):253-302

[7] Fuentes R. Epistemological Convergences and Divergences in and for the Study of Communication. Puerto Rico: VII Seminaries Research ALAIC; 2015

[8] Cropley A. In praise of convergent thinking. Creativity Research Journal. 2006;**18**(3):3404

[9] Wen Huei C. Divergent and convergent thinking projects in interdisciplinary studies. The Online Journal of New Horizons in Education. 2015;**5**(I):1

[10] Foster C. The Convergent-Divergent Model. Journal of International Society for Design and Development in Education. 2015;**2**(8):28

Chapter 4

Personalizing Course Design, Build and Delivery Using PLErify

Maria Lorna A. Kunnath

Abstract

The Course-Building technique called PLErify was developed by the author in response to the emerging roles of university faculty in the technology-driven teaching with the rising popularity of AI and deep learning. Topics that support personalized teaching and learning using technology to make it more efficient, more effective and more pragmatic. Early attempts at pedagogy and trends that pushed the personalization movement are explained. The progress of the project in a Web App format is detailed focusing on a faculty building a sample hybrid course planned for a course offering of a framework of digital resources within the app in a technology-rich smart classroom. The PLErify course-building Template is explained with methodologies to add content to it in various ways with suggestions to insert multimodal techniques, e.g., Augmented Reality, Virtual Reality and Simulation, however applicable, alongside numerical data-science-supported technologies that will comprise the most part of course presentation technique. A portion of a full course will be demonstrated using PLErify with an accompanying Course Evaluation for Professors to mull to prepare for course redesign current to improve next year's offering of same course.

Keywords: web application, digital resources, personal learning environment, PLErify, course development, MOOC, didactic, AI

1. Introduction

In a digital society, every aspect of our daily lives is interconnected and each person has an identity that is solely one's own that is encrypted and authenticable by a system. That identity allows you to interactively access multiple parts of any platform to perform actions and obtain something as a result. In an ideal version of a digital society, we humans are interconnected as citizens (*e-government*) and members of various groups (private) and afforded rights and privileges accordingly. We see micro versions of the workings of a digital society in Big Tech such as Facebook, LinkedIn, Twitter, Instagram, and Google with each connectivity model functioning according to predetermined business model.

Our ever-evolving digital society intertwines the roles of humans and robots in institutions and industries. These roles keep changing as technology advances to near capability of humans through artificial intelligence and deep learning permeating the deepest trenches of every industry, not excluding higher education. In higher education, it is hardly noticeable that the roles of main players (*faculty and instructors*) as primary owners, designers, and deliverers of their own courses for live instruction need to step up and adapt more aggressively more so than any other group. Not to be confused with purely online learning, tech-driven courses go beyond use of learning management system, LMS. PLErify use of private server, resource, and tool-based application is one such solution and discussed below.

2. Emerging faculty role in an AI-driven scenario.

> *"If we teach today's students as we taught yesterday's, we rob them of tomorrow".*
>
> *John Dewey*

An American philosopher and educator, John Dewey (1859–1952) gave a very powerful quote with a whole new meaning that is truer now than in his time. Truer now because the educational methods we now deal with goes beyond the chalkboard, goes beyond talking in front of students, and goes beyond doing projects in isolation using pen and paper. Not completely discounting the power and value of note-taking using pen and paper and would not advise against the method, it is important to recognize the presence of computational tools being used as part of current teaching methods he would have never imagined would exist today.

Undeniably, educational technology tools ushered the transition from passive (*sit, listen, take notes*) to active (*interactive "constructivist" learning*) with students defining their knowledge accumulation, construction, and learning pathways. Deep learning, Machine Learning, Big Data and Internet of Things (*IoT*), and artificial intelligence disrupting education in more pervasive ways have no specific timeframe. While innovation and adaptation slowly chip away traditional education system, the idea of faculty being put aside with little to no role in designing, offering, syndicating, and delivering courses (*aptly described as course massification*) is in fact a repelling thought. In universities however, few scenarios must play out to avoid a scenario where the machine decides and controls. The professor must play the central role but in an enhanced strategic [1] and impactful way. They (*professors*) must assume leadership roles to formulate actionable changes but be cognizant and fully prepared to face added responsibilities from programming robot consciousness to designing robotized courses through aggregating and updating content, that is, build virtual (*academic versions of Alexa-like*) robot assistants [2], automated pulling of content from a variety of resources. These new robot-driven challenges in higher education are succinctly described as follows.

2.1 Robots and professors for efficient teaching

Widely practiced in Japan, Korea, Taiwan, Singapore, and China, are robots (built in the likeness of a professor/researcher), robot applications, and robots programmed to co-teach/co-research juggling the myriad roles of the human instructor. Other foreseen creative uses of these robots involve individualizing attention to each student, thereby ensuring progress, remediation, and success (knowledgebase-driven virtual assistants). Missing in those possible roles are robots that build online courses for professors based on didactic teaching styles and student learning styles all utilizing high integrity knowledge bases with optimum performance. Past attempts at course development using course sequencing [3], adaptive learning paths [4], computational teaching, and participatory teaching [5] can inspire new innovations in this area. One deep-learn course building technique is an AI-based course aggregator (*software-based*) which pulls different curriculum (using Big Data) of the same course from many places/universities and then gets stored to a central location where students get to pick and choose a course program of study. These new courses would be up-to-date with new data that include recent developments in the discipline. Since my focus on PLErify is to assist the instructor, the adaptive learning concept for learners based on any learner model (*cognitive, behaviorist, and mental models*) has been intentionally skipped.

2.2 The MOOC-as-course augmentation for faculty and as resource for PLErify teaching learning

In AI age, faculty must face their new roles as programmer and owner/builder of learning environments of their courses that they must update per semester. Professors who remain indifferent in the new reality of a virtualized higher education vis-a-vis their expanded roles and new responsibilities will face major challenges as industry-driven automation-driven AI persistently seek dormant or stagnant unchanging areas to automate and simplify. A faculty [6] from Scotland recalls his very productive sabbatical spent at Google (*Big Tech and AI-driven*) in Silicon Valley. In that sabbatical, this academically trained faculty learned and eventually re-adapted his purely academic mindset (coding) to meld with that of a practitioner's mindset and started building coding projects that work in real life. Faculty (*particularly those in the sciences and engineering*) can follow Barker's example and come out technologically empowered fusing the academic with the practical real world and impart the same mentality to its students, that is, college to career.

Regardless, MOOC courses will continue to be made available online to anybody for free, or at a minimal cost. Boosting acceleration of acceptance by universities globally, MOOC continues an upward evolution toward a better practical higher education option for both career (*skills training or mastery certifications system*) and advanced degrees (*bachelors, masters, and doctorate*). Also, with student advancement (*and their interest in mind*) and ease of teaching for instructors as prime motivators for MOOC development, there is truly so much about the MOOC system to be appreciated that will make higher education leaner, more efficient, very current, and very affordable.

Another very encouragingly useful aspect of MOOC that is only now being realized is, they add to the personalization of learning both as a teaching resource for instructors and as an inexpensive way for students to obtain advanced degrees and lastly but more importantly to upgrade skills of work professionals. MOOC business model is being revamped and, evolving toward a more profitable version, thereby offering fee-based enrollments where certification of completion is a student's objective. The free aspect of the MOOC model however can be used by all faculty as another teaching tool. For example, professors can require students to take the MOOC version of their course offered by other universities (*now estimated at 900*) shown in **Figure 1**, before students take the real course. In a PLErify state, courses already in MOOC database can be treated as a required mastery before registering and enrolling in the equivalent actual university traditional course or program unloading faculty of heavy teaching. Some tedious portions of the course can be bypassed having mastered it beforehand through MOOC.

While debates and experimentation continue to grow in artificial intelligence, PLErify App (2007) remains a precursor to the above scenarios. Even though the core of educational technology research centers on academic applications, academe, ironically remains the most resistant and the slowest to adapt to a scenario of AI, Big Data and IoT which when taken as a group suddenly changes the course delivery game. Groups in the tech industry persistently hint at a future without a human teacher and professor, as computer scientists now and then flirt with the idea of adding consciousness to a computer. At this time though, a robot cannot actually augment human cognitive and emotional capabilities through what they claim as smarter machines currently experimented in other industries (automobile industry). I would simply and safely assume that use of virtual robot assistant is an easy spillover for use in higher education [2].

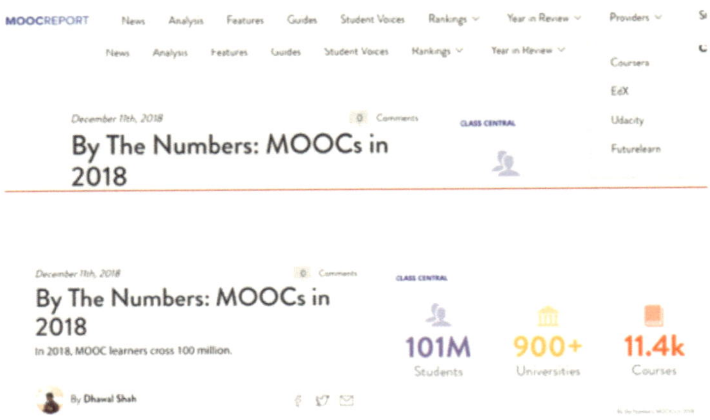

Figure 1.
The hard-to-ignore breadth and reach of MOOC globally.

It is best to speculate that whatever happens in the corporate industry will, in some form happen in the education industry. The digital society interconnects everything, from machines and app to the software/hardware; from knowledgebase to users; from different variety and degrees of transactional computing; from the teachers to the students; from the businesses to the consumers; from the students to the universities; from the faculty to the students to the universities; and finally, from the ordinary users to everything which can occur via our desktops and our handhelds. Apocalyptic ideas have been flouted at global corporate e-learning events that hint at the idea of massification to replace traditional creative teaching without a human teacher, which may appeal to select academicians who fall into the trappings of "easy teaching," that is, less classroom presence and letting the students watch video lectures and digitized .pdf files of the syllabus. Given that these handheld tools are now a normal part of everyday life blending the here, the now, and the future, a DIY culture for course building becomes inevitable. Embodied by the PLErify application (2007), the DIY mindset provides a solid training ground for ubiquitous computing vis-a-vis course building as it involves an interplay of a variety of cognitive skills combined with digital conversion of ideas into a viewable medium.

Today, 24/7 we carry our smartphones, iPad, and other handhelds also known as mini/microcomputers, more powerful than any computers built in the 1980s and the 1990s, with us and with these technologies we socialize, network, listen to music, share photos, financially transact, chat on live video, and much more, thereby doing tasks never before possible at the very same period of time educators were theorizing on learner styles, cognitive styles, etc. My own observation over this past decade is that while educators spent so much time researching learner styles and cognitive styles, they believed impacted learning, Big Tech simply went ahead and produced a plethora of handhelds and smartphones that rapidly jump-started user acquiring tech skills in turn accelerating mastery that are, fortunately, usable in both daily life and university learning but unfortunately left out those who could not keep up with the constant roll-out of new versions and models. What that phase did to each of us was it made us tech-savvy and I would argue, smarter. Now, certain tech user interactions have become ingrained for majority of us smartphone and multiple device owner and users. Majority of learner tasks to: make choices, complete learner tasks, solve problems, think about thinking (metacognition), compute, analyze have become second nature.

Indeed, technology has a very democratizing effect on its dedicated users from acquiring uniformity of skills to performing actions to obtain something back as a result; skills, which by the way, are also transferable to other domains from personal, to business, to higher education with specific attention to learners. All users get it. We can turn on the device, charge the device, download and use apps, transact, collaborate, blog, share documents, and so many other things that it is now second nature to have (*as opposed to not have*) our smart devices even while we sleep. Technology has intercepted our lives in unimaginable and remarkable ways psychologically but best of all, educationally. I must conclude that though technology tools are not advisable for use by children, technology for mature adults is an additive rather than a subtractive experience.

3. Didactic models for creative computational teaching

The timely re-entry of computational [7] tools to teach creatively befits this era of our technology-driven education. Less intervention on how students create their learning paths as they meld new learning with what they already know in working memory gives students a better grasp at how to manage their interactions and the accumulation of those interactions in a self-directed way exemplified by the constructivist didactic model used in the Virtual Mentor Project notably learning by asking LBA Project [8].

The actors on stage in the world of tech-based teaching and learning and their functions in the teaching learning equation are summarized with one infrastructure in common: connection to the Internet (**Figure 2**).

Software applications accessible on the web allow both instructors (*course makers*) and students (*users*) to manipulate course content to create, present, and store. *Learning Management Systems* (*LMS*) are prepackaged applications that act as an administrative tool to manage the online course, the students who enroll, and the professor who offers the courses. *Artificial intelligence* (AI) is the byproduct of deep learning, huge swathes of databases within a database that when meshed together gives it intelligence though within a limit, that is, you can program a robot to do certain things that will be limited to the amount of intelligence you put in it. A human is still in command and an ill-programmed robot like a biased robot gives

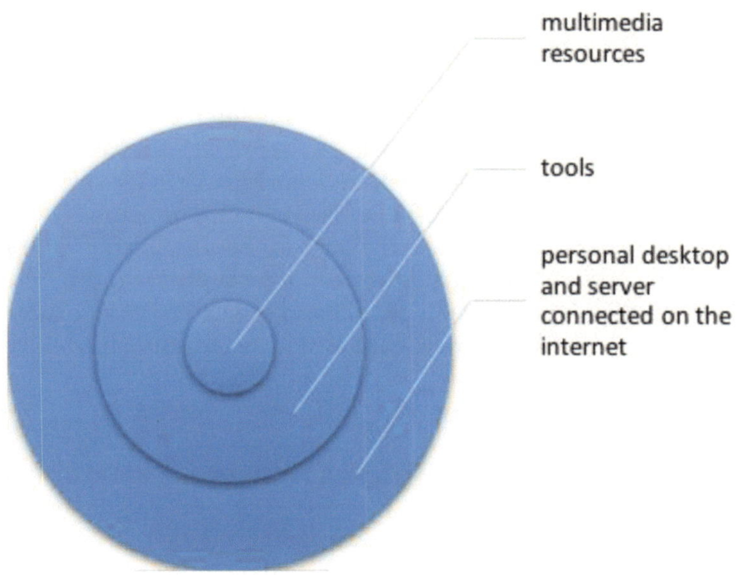

Figure 2.
Technology infrastructure to teach.

disastrous results. *Data science* is the highest application of computational ability that can detect patterns of naturally occurring events in nature, analyze it, and provide very accurate predictions and analysis in context. *Interactive rooms/walls* that began in the mid to late 1990s, but was not very successful such as the project at Stanford University and though unsuccessful at its early attempts to merge it in the teaching and learning game, has re-emerged (*Germany Applications*) with more sophisticated screens that cover the entire walls making possible the projection of visualized data comprehensible to all learners.

Augmented reality when applied has the capability to freeze, slow down or speed up, and look for patterns among piles and piles of data and of events that are intended to be analyzed and deduce from. It is very useful in the science and engineering course making in particular, though also usable in other fields when observation of actions is sought. *Virtual reality or virtual environments* are computing environments otherwise called *virtual immersions* where "users are immersed in low latency high-quality visual stimuli and spatial audio amplified by motion platforms, dispersed systems, and active/passive haptic device that allow users to fill objects with more sophisticated systems equipped with gesture recognition and voice input. Success of VE systems depend on system performance matching user expectations" [9]. *Ubiquitous computing* is computing at its finest you hardly notice it. *The Clouds* are on demand delivery model that enables the synchronized delivery of computing resources such as applications, storage, servers, networks, and services (liberating software providers of low-level IT on premise infrastructure) that allows instant scale up of various types of web services [10]. *Web services* are on-demand applications, storage, and servers and sometimes called SaaS, PaaS, and IaaS enabled mostly through the cloud. *DevOps and Containerization* [11] is a systems management technique for on-demand services performed in a bundled way as a completely packaged infrastructure (*operating system and software applications*) updateable on the fly with the least hassle.

Bonk [12] aptly describes a changed e-learning ecosystem in the past two decades and summarizes it based on three themes namely *Learner Engagement*, *Pervasiveness*, and *Customizability* using the same computing technologies for document-sharing, collaborating, software, hardware, communicating, and digital resource use. Based on learner engagement, he cites, 1—mobility, 2—visual, 3—touch sensory, 4—game-based, 5—immersive, 6—collaborative, 7—social, 8—digital and resource rich, 9—adventurous, 10—hands-on and in its pervasiveness, 11—mostly online, 12—video-based, 13—global, 14—immediate, 15—access to experts, and 16—synchronous in being customizable, e-Learning is more 17—open, 18—more blended, 19—more competency-based, and 20—ubiquitous. With its customizable quality, e-learning is 21—more blended, 22—more self-directed, 23—more competency-based, 24—more on demand, 25—more massive, 26—more modular, 27—more communal, 28—more modifiable, 29—more flipped, and 30—more personal. The "more personal" thus sets the stage for personalization (PLErify) in both teaching and learning, which lends itself appropriately to the different didactic models [13] of teaching for different types of courses whether it is intended for knowledge-building, theorizing, knowledge or skills acquisition, model-building, simulation, and argumentation.

4. General instructional design with less focus on user learning style

Instructional design for high-performance computing [5, 7, 14, 15] focuses on the principles governing working memory vis-a-vis cognitive load [16] extracting memories associated with completion of task (primary and secondary memories). Primary memory are those cognitive schemas a person acquires as a result of interacting with the environment stored in long-term memory which the secondary memory (*the cognitive schemas stored in short-term memory*) uses to solve problems though not without limits. Knowledge creation for working memory follows the principles of (a) "narrow limit of change," that is, within a small span of time, new information should be very limited in order for it to be stored in long-term memory. "Unlimited": it may be in the amount of information it can process, working memory follows the principle of being capable of spewing (b) "unlimited amount of information" for information retrieval. When that knowledge creation does not occur, the mind adjusts by following the principle of (c) "randomness as genesis" methods (the borrowing and reorganizing principle, that is, imitation, listening, reading, and social interaction) to compensate in the knowledge creation. Finally, when cognitive overload must be overcome to commit new knowledge in long-term working memory, the mind does what is called the principle of environmental organizing and linking, retrieving, and cycles through information already stored in long-term memory.

In non-scientific, non-engineering subject matter, focus on cognitive load combined with learning theories has not been exhaustively studied. Learning styles has been linked to the effective design of course materials as it affects comprehension and overall performance [17]. A person's style of learning is determined by environmental factors manifested through behavioral patterns. In my 2001 Doctoral Dissertation [15] experimental study based on learning style effect on user performance, I found out that in a matched condition, i.e., matching concrete icons with concrete learners and matching abstract icons with abstract learners resulted in better performance on recall and memory and task completion. Concrete learners performed better overall in a matched condition. There are other learning theories besides that was used (Kolb's) in my experiment and most are in the style of thinking and therefore behaving, extent of proficiency or lack of and style of responding to environmental triggers. Knowing fully well that style of learning in the AI-driven teaching will override the learning style consideration, platforms will be built mainly based on learner independence during the learning process. That is, they will determine the route, path, and speed at accumulation of knowledge and skills as they see fit. In the past, there was "adaptive learning" where the computer adjusts to the learner based on the speed of knowledge acquisition of the user and then readjusts the next set of materials based on that performance. If the previous task proved hard, the computer generates an easier task to complete and vice versa.

Though the idea that learner pathways must still be considered in designing personal learning environments, it is safe not to overly worry about learners and skip the time-consuming practice of hand-holding knowing that users have full control of their digital strategies and techniques to learn. It is both consoling and problematic at the same time: consoling because instructors would not need to look over learner's shoulders during the process of mastery, yet problematic because it now forces the instructors to be on the top of every technology used by the learner. Instructors need to possess tech skills better than students. Casual everyday users (*including the dark forces of the web*) of tech will possess mastery of technology for every intended purpose.

5. State of purely online learning

MOOCs, such as EdX, Coursera, Open CourseWare (OCW), and hybrid designs, are designed to offer free courses for poor countries (MOOC), to corporations (*Coursera* and *Udacity*), EdX (*universities*), however of late modified it to a paid model adding some validation features to address legitimacy of courses, such as granting of certificates, shortening of Master's program as the case in Georgia Tech. Student attrition remains a problem compared to regular traditional classroom model proving that regardless of convenience, students still prefer a human teacher in the classroom indicative of what history has revealed, that is, the most valuable teacher is a human. Another acute finding is that MOOC courses are not suitable for advanced courses that can only be handled by a human.

Alternatively, MOOC courses, based on a very interesting observation of Cooper and Mehran [18], have the potential utility in personalized learning in the same manner as YouTube online video courses do, that is, a place to find highly reputable learning resources for students to pre-familiarize themselves of courses they will take before they turn up at actual class lectures. This utility when applied as a "before-you-attend-a-class" feature skirts the nagging issues attached in MOOC such as validation, plagiarism, certification, and lack of richer evaluation. MOOC, in that capacity, is indeed a welcome addition to personalized learning. One monetization [19] possibility explored by MOOC concerns that of providing added validation about the student for employment which, to my mind is very interesting and closes the loop of education to career. In an interview [13] with John Hennessey by his longtime colleague Davis Patterson, John was very enthusiastic about MOOC and thought of it is a compelling solution for continuing education (skills upgrading for working professionals) with his continuing belief that Masters and PhD program will be part of MOOC and non-MOOC.

In that vein, professors in higher education institutions need to skill themselves sufficiently to be able to create a digital course only once but updated for every semester's offering. Faculty load of work is, in truth, lightened while students carry most of the load of a course, that is, reading materials, accessing mixed modal multimedia, collaborating, project work, homework, assignments, critiquing, mid-term exams, and final exams. In the PLErify platform, AI tools, in the research (*Research Resources and Libraries Semantic Knowledgebase*), analytics (*Quantifying/Analytical engines*), authoring (*Multimedia and Multimodal*), and learning management systems (*Course Delivery Student Access Point*) are placed in a private server for the instructor where he is given the freedom to extract all sorts of information for his course and capture and store those information in an organized cataloged file directory (*on both desktop and the Private server*) purely for his own access **Figure 6**.

6. PLErify course design and future AI prospects

PLErify components that are visually depicted in **Figure 3** are as follows.

Research is a set of live databases to extract content through digital libraries for download of scholarly materials, quantitative and qualitative data. *Analytics* is a set of measurement tools to analyze datasets and extract visualization files for inclusion in the instructor's course curriculum. *Authoring tools* are a set of applications to convert text to viewable mixed media and mix it up for immediate playback as one full course. *Learning management system* (*LMS*) is described in my paper [1] "Virtualized Higher Education" as the course's administrative tool performing the tedious tasks as the container that holds a course or courses. An LMS is an application that typically contains the following: administrative features for managing content: content uploading/downloading; calendar and scheduling instructor timings; student administration; faculty/student communication tool; conferencing; and homework/project submission and grading. These processes and activities as shown in **Figure 4** are time-consuming and repetitive some of which can be made efficient through automation. Decisions to automate parts of the PLErify platform depends on the instructor who (*after updating his skills in AI, deep learning, virtual assistants, and robots*) can select which activities to augment as shown in **Figure 7**.

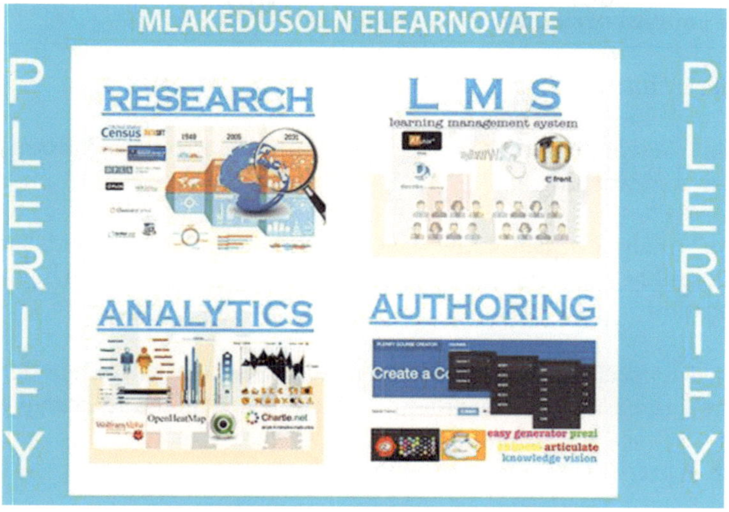

Figure 3.
The PLErify application toolset desktop version (http://elearnovate.com/plerify).

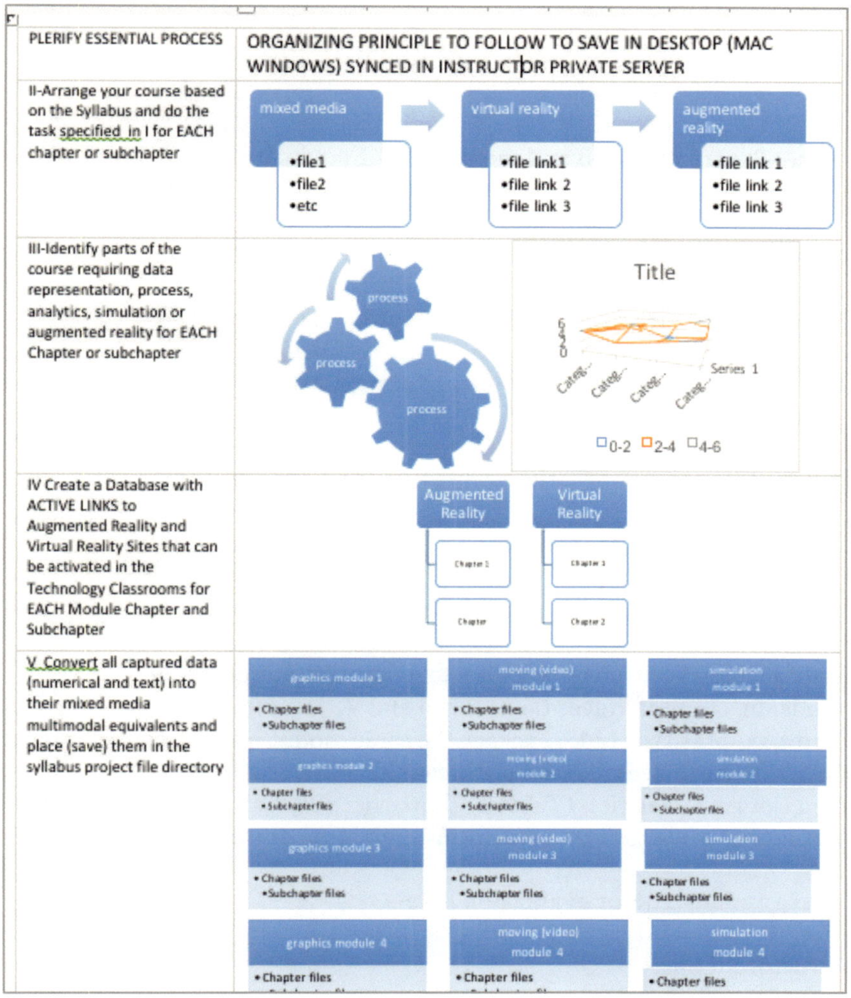

Figure 4.
PLErify processes and organizing storage principles for the Mac and Windows.

6.1 Sample course curriculum on learning styles

- Module I: Timeline Origins and Theorists
    - Step 1 Research: gather literature on early theorists (background, philosophies, and teachings)
    - Step 2 Analytics: quantitative data from research associated with background, philosophies, teachings, and applications of learning theory in higher education
    - Step 3: extract visuals and moving media equivalent of content derived from steps 1 and 2 and combine them to illustrate concepts and examples
    - Step 4: prepare the interactive module summary for Part 1.
- Module II: Belief Systems of LS (Cognition, Behaviorism, Environment)
    - Step 1: gather literature on LS relative to cognition, behaviorism, and environment
    - Step 2: extract quantitative/qualitative data on LS relative to cognition, behaviorism, and environment
    - Step 3: extract visuals and moving media equivalent of content derived from steps 1 and 2 and combine them to illustrate concepts and examples
    - Step 4: prepare interactive module summary for Part II.
- Module III: Higher Education Applications of Learning Styles
    - Step 1: gather research on higher education use of learning styles (projects successful or unsuccessful)
    - Step 2: extract quantitative or qualitative info based on above.
    - Step 3: visuals and moving media from steps 1 and 2.
    - Step 4: prepare interactive module summary for Part III.

Identify simulation videos or games to illustrate LS application.

Include the URL's of videos (simulation and VR) within the course before packaging it for export to the LMS. Package the three modules as one course and export it to the LMS enroll students taking the course.

Add a Course Evaluation (**Table 1**) showing a generic form freely available online at https://www.jotform.com/form-templates/course-evaluation-form-3.

6.2 Course and instructor evaluation form

Instructor's Name.
Course Description.
Course Number Date-Month-Day Year.
Please evaluate honestly.

	Excellent	Very Good	Good	Fair	Poor	Very Poor
The course as a whole was:						
The course content was:						
The instructor's contribution to the course was:						
The instructor's effectiveness in teaching the subject matter was:						
Course Organization was:						
Clarity of instructor's voice was:						
Explanations by instructor were:						
Instructor's use of examples and illustrations was:						
Quality of questions or problems raised by the instructor was:						
Student's confidence in instructor's knowledge was:						
Instructor's enthusiasm was:						
Encouragement giving students to express themselves was:						
Answers to student questions were:						
Availability of extra help when needed was:						
Use of class time was:						
Instructor's interest in student's progress was:						
Amount you learned was:						
Relevance of course content was:						
Grading techniques were:						
Reasonableness of assigned work was:						
Clarity of student requirements was:						
Intellectual challenge was:						

Table 1.
Generic course evaluation form that can be modified.

6.3 Student participation

The amount of effort you put into this course was:

Excellent Very Good Good Fair Poor Very Poor.

On average, how many hours a week did you spend on this course (in and out of class)?

0–2 2–5 6–10 11–14 15 Up.

What grade do you expect in this course?

A (4.5–5.0) B (3.5–4.4) C (2.5–3.4) D (1.7–2.4).

This course is best described as:

Major Minor A distribution requirement A program requirement Prerequisite Other.

Every e-learning course is organized into modules shown in **Figure 5**. To populate content, the method is quite straightforward starting with Module 1 but not necessarily following a linear process, that is an instructor can jump from Module 1 to other modules in no particular order depending on how they interconnect topics and ideas.

Click Module 1. Module 1 will load chapters. In the edit mode, you can replace the content with your content. Chapter 1's format is repeated for Chapters 2–4. You can replace the content as your syllabus progresses. In each chapter, you can include datasets (from the research toolset analyzed with results presented).

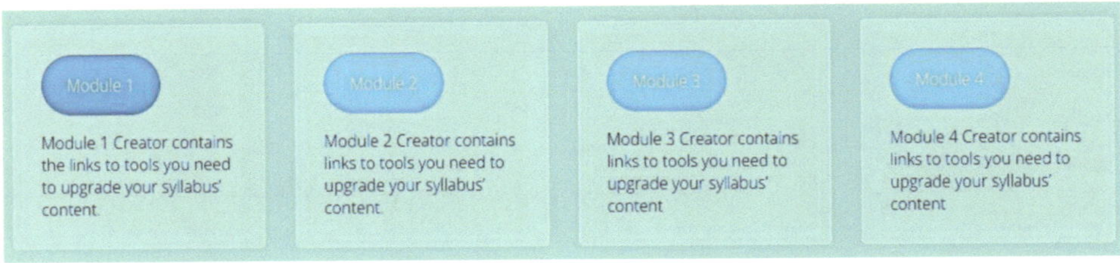

Figure 5.
PLErify course creation in modules.

These analyses of presented data, or sample data can be saved in database readable format backed up in instructor's private server and desktop for inclusion in the digital course. The content of the modules is managed as shown in **Figure 6**. For example, in a digital course on Learning Theories, an instructor will find the timeline data to present the history of the early to modern learning theorists. This timeline tool in the PLErify App can be dramatized through an augmented reality historical film on the significance of each era and how it influenced education at different times in the history of the modern world.

Personal learning environments or expert systems as it is sometimes called is disruptive enough to education due to its "lean to use automation." Any AI application is still limited in capability where human skills of negotiation, detection, mobilization, and understanding of power and trust (*much like the gut instinct humans have to sense danger and change*) is required. In other words, regardless of whether it is continually built to be smarter and smarter and contradicted by Krakovsky [20] who has a more optimistic view, AI as predicted may not develop a true "sense of self." Her research to a certain extent can be useful for building the intelligent robot as instructor assistants that will be tasked to meet students, answer course-related concerns, substitute the human instructor who is on research travel, and track student progress as described in Section 2.

Assigning automation features shown in **Figure 7** in PLErify in the next 5–10 years will center on course preparation, in converting a simple text to something more graphic or visual, combining the visuals into a more powerful single visual based on context, capturing real live data from a source known only to the instructor, citing the link of that source in the course materials, mastery in the use of sophisticated tech-enhanced classroom, synching course presentation of materials with the tools in the smart tech-enhanced classroom, and automating tasks in use of the LMS.

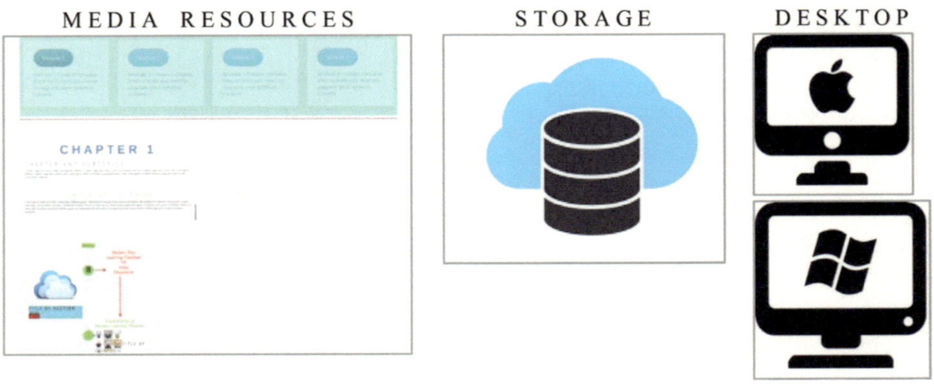

Figure 6.
Course module management.

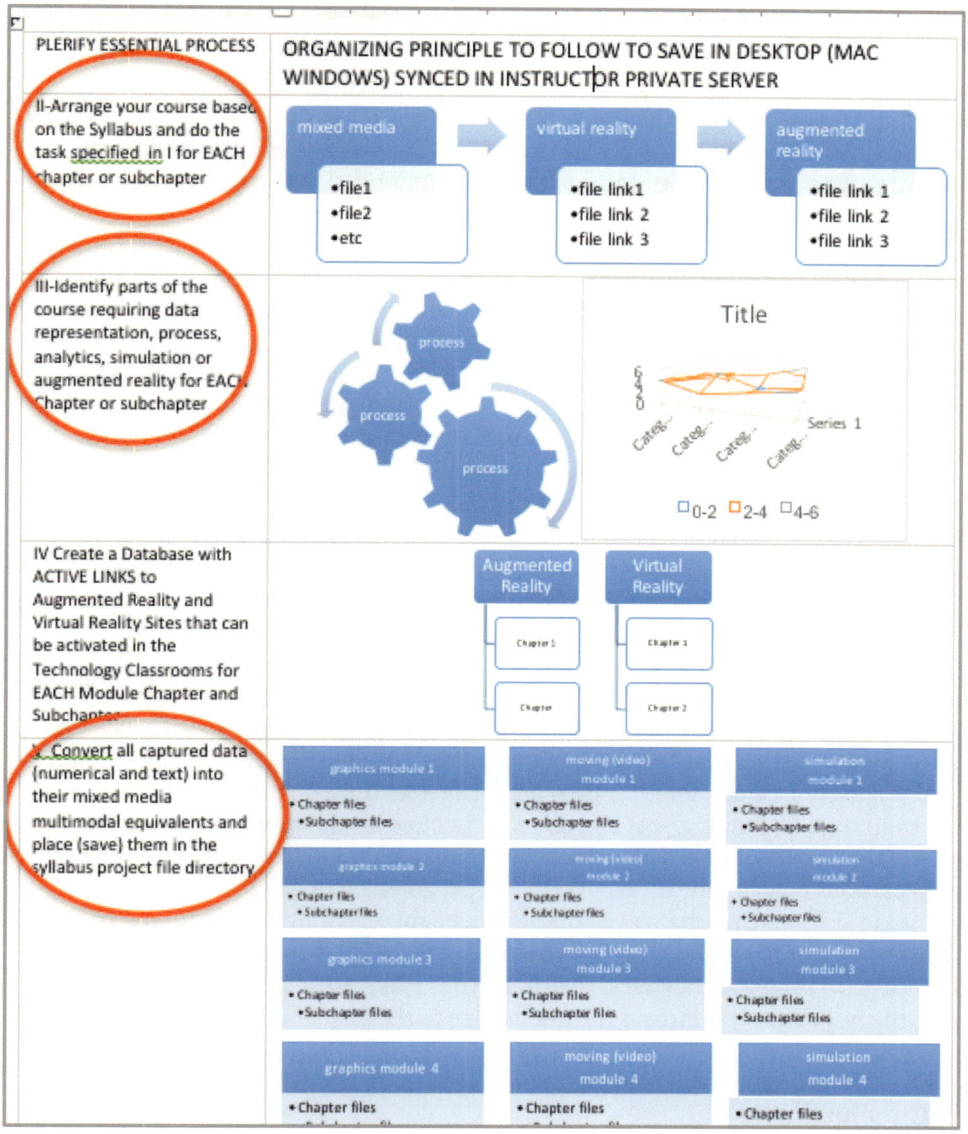

Figure 7.
Areas in PLErify that may be automated highlighted in red circle.

7. Reflections on the profession vis-a-vis digital society

We can entrust the ability to recognize learner styles, learner abilities, comprehension and understanding in Artificial Intelligence (AI) as it continues its ascent towards enhanced intelligence in almost all facets of our digital life in this case Higher Education and Course building. In that token, Instructor (*as Designers*) and Technologists can look into Chris Stary's [13] research on the Scholion Project as one guide looking into the design and implementation of his constructivist-based project that aids learners taking into account learners' mental models, cognition and metacognition.

In this second half of this decade, AI's recognition capability has gone far beyond its early beginnings that it is now termed the Age of the Machine. Much similar to Elon Musk's Tesla, the machine can now build other machines. Thinking about this new reality in education also means the teaching and learning can now rid of a lot of the mundane tasks in: course creation (*by instructors*), course management (*by LMS's*), course participation (students), grading and course evaluation methodology, and issuing valid certification. The machine age that will make our transition to a fully digital society brings with it a whole slew of new realities as well for the professional. The expectations from each of us has become compounded to an extent that a degree certificate attached to an individual that requires skills are also updated be it technical, socialization, community, or connectivity [21].

Professional degree certificates (*undergraduate and graduate*) now also comes with it the responsibilities of skills upgrading on a continuing basis; of being in communities attached to that profession; of being in the know in those communities; of being socialized in the platforms of Facebook, Twitter, Instagram, and LinkedIn to not just being there but being a part of the larger whole comprised of billions of people in the world. To think about it in terms of how AI responds to that changed perspective is to assume that all these interactions are captured by the AI machine and adds it in terms of quantifiable data, making new assumptions about the added data in effect, understanding ourselves more from that newly captured information producing an updated profile of a person's learners style based on captured cognitive functions and user action decisions as opposed to self-reported learner style [22].

Schneider [21] points out that through data analytics, these learner characteristics can be extracted automatically from user's ongoing interaction to perform a variety of transactions (*from finding a route using GPS (geographical positioning system) or learning online to online banking transaction*) in day-to-day lives. When these patterns of user interaction are validated, a method called "user nudging" could be used much like adaptive computing that "adjusts to serving user tasks based on prior action," that is, nudging would guide user actions by giving choices through prompts.

Denning [23] summarizes it brilliantly that to truly survive in the age of machines where the knowledge worker conducts work on highly intelligent machines, new expectations come to the fore that requires pragmatism in belongingness, ever adapting skillsets that changes as the system changes, community building based on chosen areas of belongingness (professional, leisure, or recreational), and last but not least, willingness to mentor, to display your skills to the person that needs it so that the next learner improves the knowledge to the next and so on and so forth.

8. Conclusion

8.1 Security concerns with respect to PLErify, MOOC, and tech tools

Security breaches from China, North/South Korea, and Russia are a threat to our tech-enabled life. These countries' very advanced cyber-surveillance and intrusion system have penetrated US cyber defense system potentially undoing major education technology advancements. The industry needs to come up with a very strong authentication system as well as cyber-blocking mechanisms beyond the obvious firewalls.

Without a strong cyber security strategy attached to all these tech innovations, any attempts at technologizing higher education would face enormous challenges. China's breaches covered the entire hardware/software and telecommunication ecosystem (home routers included) baffling Europe, the US, and Australia. A solution that has been proposed is virtualization and containerization. If virtualization and containerization provides a guarantee for the safety and security of cumulative progress and strides made in the education sector and if we are willing to adapt to rapid changes demanded of us as educators, faculty, and students, then the future will certainly be bright.

Acknowledgements

After incorporating additional critique (style, formatting, and punctuation) from my children who are both writers in their own right, I finally produced my chapter contribution to the book 2019 Engineering Design and Innovation Methods at IntechOpen. I take this opportunity to convey my deep heartfelt appreciation to IntechOpen's generous partial publication waiver which assisted me in pursuing further publication of my manuscript with them. Thank you.

For being my constant inspiration and for whom I am constantly reminded that we must perform our responsibility as trustful stewards of technology for the future of the young generation and the forthcoming ones thereafter, I dedicate this work to my children Avinash A. Kunnath (AB Mathematics, University of California Berkeley, 2008) and Ameeta A. Kunnath (BEngg Structures, University of California San Diego, 2015) who, along with members of their generation, will continue the work in their chosen careers and professions to improve various facets of our digital society we are all part of as technology continuously shapes and impacts our modern lives by the minute.

Author details

Maria Lorna A. Kunnath
Owner, CEO, Founder of MLAKEDUSOLN ELEARNOVATE, Davis, California, USA

*Address all correspondence to: mlkunnath@elearnovate.com

References

[1] Kunnath MLA, Virtualized higher education: Where e-learning trends and new faculty roles converge through personalization. In: International Conference on Information Society; IEEE UK; Dublin, Ireland. 2016. pp. 109-115

[2] White R. Skill. Discovery in virtual assistants. CACM. Nov 2018;**61**(11):106-113. DOI: 10.1145/3185336

[3] Brusilovsky P, Vassileva J. Course sequencing techniques for large scale web-based education. Journal of Continuing Education and Lifelong Learning. 2003;**13**(1/2):75-93

[4] Alzahrahni et al. Towards personalized and adaptive learning paths in immersive educational environments. In: Immersive Environments. London: Kings College; 2013

[5] Hamilton E, Owens A. Computational Teaching and Participatory Teaching as Pathways to Personalized Learning

[6] Barker A. Viewpoint an academic's observations from a sabbatical at Google. How experiences gained in industry can improve academic research and teaching. CACM. Sep 2018;**61**(9):31-33. DOI: 10.1145/3177748

[7] Houstis E, Gallopoulos E, Bramley R, Rice J. Problem solving environments for computational science. IEEE Computation Science and Engineering. Jul-Sep 1997:18-21

[8] Mishra P, Hershey K. Etiquette and the design of educational technology. CACM. 2004;**47**(4). DOI: 10.1145/975817.875843

[9] Whitton M. Making virtual environments compelling. CACM. 2003;**46**(7):40-46. DOI: 10.1145/792704.792728

[10] Goutas L, Sutanto J, Aldarbesti H. The building blocks of a cloud strategy: Evidence from three SaaS providers. CACM. 2016;**59**(1):90-97. DOI: 10.1145/2756545

[11] Roche J. Adopting devops practices in quality assurance. CACM. 2013;**56**(11):38-43. DOI: 10.1145/2524713.2524721

[12] Bonk C. What is the state of e-learning? Reflections on 30 ways learning is changing. Flanz Journal of Open, Flexible and Distance Learning. 2016;**20**(2):6-20

[13] Stary C. Didactic models as design representations. In Jacko, J.A. (ED). Human computer interaction part IV. HCII 2009. LNCS. 2009;**5613**:225-235

[14] Houstis EN et al. MPSE: Multidisciplinary Problem Solving Environments. Department of Computer Science Technical Reports, Purdue University e-pubs. Report number 95-047. July 1995. pp. 1-9

[15] Beng Lee C, Leppink J. Instructional Design Principles for High-Stakes Problem-Solving Environments. Singapore: Springer Nature; 2019

[16] Kunnath ML. Effect of Pictorial Icon Interface on User Learner Performance [Dissertation]. Orlando, Florida: University of Central Florida; 2001

[17] Cooper S, Mehran S. Viewpoints education reflections on Stanford MOOCs. CACM. 2013;**56**(2):28-29. DOI: 10.1145/2408776.2408787

[18] Dellarocas C, Van Alstyne M. Viewpoint economic and business dimensions money models for MOOCS. CACM. 2013;**56**(8):25-28. DOI: 10.1145/2492007.2492017

[19] Patterson D. An interview with Stanford University president John Hennessy. Stanford University professor discusses his academic and industry experiences with UC Berkeley's CS Professor David Patterson. CACM. 2016;**59**(3):40-45. DOI: 10.1145/2880222

[20] Krakovsky M. Artificial (emotional) intelligence. CACM. 2018;**61**(4):18-19. DOI: 10.1145/3185 521

[21] Schneider C, Weinman M, Brocke JH. Digital nudging: Guiding online user choices through interface design. Communications of the ACM. 2018;**61**(7):67-73

[22] Simonite T. Teaching machines to understand us. MIT Technology Review. 2015;**118**(5):71-77

[23] Denning P. The profession of IT automated education and the professional. CACM. 2015;**59**(9):34-36. DOI: 10.1145/2804248

Chapter 5

Mechanical Engineering Design: Going over the Analysis-Synthesis Mountain to Seed Creativity

Kant Kanyarusoke

Abstract

This chapter advocates and exemplifies a change in delivering mechanical engineering design (MED) to undergraduate students. It looks at, and critiques the current delivery mode which treats MED as an extension of Natural and Engineering Science, through its bias for analysis of existing systems. It is argued that even though students' innovativeness might be getting slightly enhanced, their creativity is stunted by the mode. So, is their understanding of how machines evolve from human needs, and of how non science related issues affect the evolution. A new teaching approach which attempts to align student thinking and learning activities with what exists in industrial MED is suggested. In this approach, human needs drive engineering problem formulation, which in turn, precipitates a synthesis of machines, mechanisms and constituent elements to satisfy the needs in a regulated environment. The regulation obeys laws of science but is mostly, 'Humanities'—constrained. Creativity and innovation case studies are given, and it is shown how new machines can come into existence in the course of learning MED. This would be difficult in the current delivery mode. The new mode, of synthesis followed by iterative analysis, helps students build self-confidence and prepares them better for industry.

Keywords: analysis, creativity, innovation, mechanical engineering design, synthesis

1. Introduction

Mechanical engineering design (MED) deals with conceptualising, planning, optimising and communicating mechanical systems to do specific tasks [1]. The tasks are meant to satisfy specific needs as desired by Man. In a most general form therefore, human needs satisfaction, requiring tasks to be done in a mechanised way, are the primary drivers of MED [2]. These needs could be anything from physical, such as moving between places 'X' and 'Y', to thermal comfort as in air-conditioning, to egoistic and futuristic as in imagining being part of a generation that sends human species out of the solar system, etc. It is evident that these need-drivers can be diverse and very complex: sometimes, they may neither be directly related to ordinary science, nor to normal expressions of art. Yet, the systems which have to do the tasks are physical. They are regulated by laws of Physics and Mathematics—whether known, or yet to be discovered. Moreover, they are expressed in artistic form to appeal to potential users and handlers. This marks the first hurdle in MED: to relate the obscure needs to as yet, inexistent systems. At engineering student level, this is perhaps the greatest challenge. We shall shortly see why this chapter symbolically refers to it as a 'mountain'.

Physically, mechanical systems consist of materials—shaped, sized and connected in such a way that energy can be input at certain points to cause desirable changes at other points within the system [3]. 'Desirable' here, means the 'changes' at those other points positively contribute to satisfaction of needs. The simplest identifiable material in the system is called a machine element. The contribution in most systems is through several groups of connections of elements, called mechanisms, which in turn are also interconnected to form the total system, or machine. Therefore, MED has to consider selection of materials for the elements, sources of input energy, and transformations of this energy within the machines being designed. Prior education and training of mechanical engineering students tends to prepare them quite well for this part of MED. This is especially so, because MED, as a subject, is normally taught later in their studies, after they have done a fair amount of engineering science subject modules. Hence, many undergraduate MED curricula tend to focus on design of individual machine elements, as typified in Refs. [4-6], and in text books [7–9]. Necessary and convenient as this may be, it creates a mental comfort zone for students that tends to further disable them from connecting the obscure human needs to the very machine elements they may be studying. They are in one valley of comfort, while the needs are in another. An invisible mountain separates the two valleys. How do we make that mountain visible—and how do we help students ascend, and then descend it? Those are the two questions addressed in this chapter.

MED is not simply the identification of needs, and inventing or conceptualising machines to satisfy those needs. In a world of ever increasing scarcity of both materials and readily exploitable energy resources, and where many other engineering designers are competing to satisfy the same needs—possibly in different ways, MED has to include a consideration of alternative and/or complementary designs. The alternatives have to be compared with, and contrasted against, each other on well-defined criteria. Complementary designs may be necessary to extend market outreach. To the extent that these comparisons and contrasts can be modelled mathematically, and analytical optimisation procedures carried out, engineering students have little difficulty in this area. However, careful consideration of needs, gives rise to two questions. One is on extents to which the needs are likely to be met; the other is on how infrequently and for how long in a given period, they are not likely to be met. The first of these concerns, contributes to quality of the design. The second leads to reliability. These two areas are probabilistic and are less familiar to students than the physical or 'functionality' part of MED. Along with them, come others characterised by chaos. These include marketability, effects on and by the environment, etc. All these issues (Functionality, Quality, Reliability, Marketability, Safety and Environmental, etc.) have to be planned for in the design. Finally, the design has to be persuasively communicated.

The endpoint of MED has traditionally been sets of detailed engineering drawings [10]. Today however, it may, in addition include: a set of simulations and their results, a working physical model, a working prototype and a series of oral and written presentations. This author considers that as much as possible, mechanical engineering students should not be let to end designs at drawings alone. This is because at their stage of professional development, they have not yet mustered sufficient insights on manufacturing and assembly processes to give error-free manufacturing drawings for workshop personnel to make and assemble satisfactory machines. The author finds that—requiring and guiding them to translate their drawings into models or working prototypes, greatly helps them improve their overall design and manufacturing abilities. More importantly, drawings, and simulations, do not produce the same level of satisfaction and self-confidence building as a finished working model or a prototype. One case in this chapter illustrates the principle of ending with a working prototype while the other, builds on a similarly finished student project.

The remainder of the chapter is therefore arranged as follows: we begin with a quick description of engineering analysis, to which, most MED students and practitioners are used, and in which, they easily find a comfort zone. Then, as a point of departure, we present a sample of industry design processes as reported in the literature. In Section 3, we present two cases: one is by the author, on design evolution of a hydro mechanism he invented in 2015. The second is by a physically challenged student, building on previous work. The originality and contribution of this chapter is in demonstrating an alternative method of delivering MED courses in order to quicken nurturing of innovation and creativity among mechanical engineering undergraduates. In the conclusion section, we summarise the differences between the two delivery approaches.

2. A brief on engineering design processes and methods

In this section, we first present the current state of handling MED at undergraduate level. We show it as being biased towards engineering analysis, rather than to the more desirable engineering synthesis. In the second and third subsections, we turn to how engineers in industry do MED. In one, we debrief the reader on processes, while in the other, we describe recorded methods.

2.1 Engineering analysis

Engineering analysis works on an existing system, which may be real or virtual in form. It applies already known laws of science and engineering to check both functionality and feasibility—if virtual. By functionality is meant, a 'YES' to the question: does this system do what it is intended to do? Feasibility means—a high (acceptable—in the circumstances) probability that the imagined system can be made and that, after then, it will be functional. The applicable laws of science consist of the virtually 'inviolable' and universal principles (within limits of present knowledge) of Physics and Chemistry, usually, but not always, as explained by Mathematics. In mechanical and chemical engineering, for example, laws of motion and of thermodynamics are good examples [11, 12]. So are those of electric and magnetic circuitry, and of logic systems in electrical and electronic engineering [13], etc. The second group—i.e., of engineering—however, are not necessarily inviolable. Nor do they have to be universal. They are practice—based. These engineering practice principles distinguish the engineering professional from the physical scientist in ways similar to how a medical doctor is different from a biologist, or an agriculturalist from a botanist. In mechanical engineering, such principles include those of making parts of the system; joining and assembling into subsystems, and finally into the finished system. Then, there are principles related to system usage, e.g., legality, cost, safety, security and environmental impacts, etc. It is clear that both the making and usage principles can vary from place to place and with era, depending on levels of development and acceptability in the societies where the systems are to be made or used.

In countries like South Africa, Botswana and Kenya, where pre-university education consists of 12 years at 'primary' and 'secondary' levels [14–16], University engineering curricula tend to start off with a consolidation of physical science and mathematical principles, and are then, in mechanical engineering, followed by an introduction to 'making' principles. In others like Nigeria, Uganda and Zimbabwe, the consolidation starts before admission to university in a so called 'Advanced' level of education [17–19]. Here, the mechanical engineering student starts off at a slightly higher level, is introduced to engineering communication, and to other essential branches like electrical and materials engineering in addition to some of the 'making' principles. MED in either case is introduced later, with analysis of virtual systems.

Even when real systems like engines, motor vehicles, home use machines, etc. are available, they are rarely analysed as whole systems because universities tend to compartmentalise knowledge. For example, in the case of a car engine, the student would have to draw on learnings from 'experts' in Thermodynamics, Mechanics of Machines, Fluid Mechanics, Materials & Manufacturing Engineering, Environmental Science, Electrical/Electronics, etc. These 'experts' would have taught the respective 'knowledge compartments' most generally, often, not even mentioning the engine. For the average student, integration of these 'compartments' in MED can be a very difficult first step to make, up the symbolic mountain mentioned earlier.

The usage principles occasionally come superficially in some final year projects. Even then however, the current approach to MED fails to motivate creativity in part, because it deals with already existing systems, whether imaginary or not. We can accept that it can lead to innovation as when an existing system is modified substantially to perform the same function 'better' or to perform others it originally was not intended for. We still note however, that limitations can be imposed by an insufficient grasp of the usage principles. To summarise therefore: to the extent that current treatment of MED at universities is theoretical analysis—driven, relying on existing systems and with limited concern for usage, it stunts both innovation and creativity. The intent of this chapter is to advocate and demonstrate a reversal of that approach, and align it with the practice in industry so that on one hand, students appreciate MED better, and on the other, they can find it easier to settle in industrial practice after they leave campus. **Figure 1** shows the two approaches, side by side.

2.2 Engineering design processes

In industrial practice, design approaches have been formalised to ensure as much detail on user requirements and on limiting constraints are taken care of, to get as cost effective (or profitable) a safe and marketable product as can be achieved. **Figure 2** shows some of the recommended processes in the literature. They all have the following characteristics [20–26]:

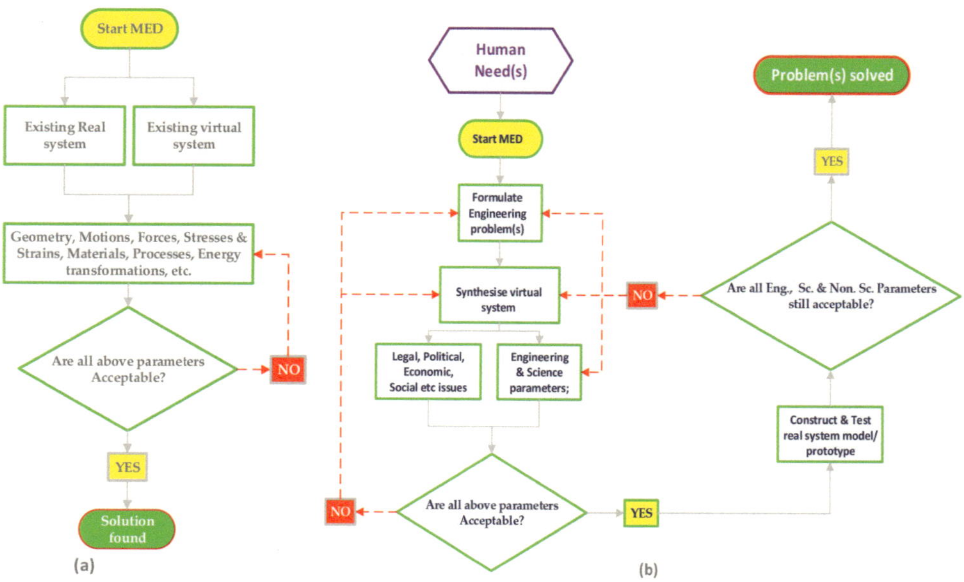

Figure 1.
(a) Current and (b) proposed teaching and learning MED approaches.

DOI: http://dx.doi.org/10.5772/intechopen.85174

- They start with a 'needs' identification, followed by problem formulation. This means: 'needs', and hence usage principles—but not analysis, drive the process.

- They involve many solutions to the same problem. This means: in different circumstances, any other solution could be appropriate—much unlike in the current class room analysis driven approach.

- They are highly iterative. This indicates incorporation of a trial and error methodology, quite unfamiliar to, and unappreciated by engineering students.

- In the cyclic processes, there is no definite endpoint. The working product at step 8 is to be continuously improved upon, depending on emerging constraints and needs.

2.3 Engineering design methods

Nigel Cross [21] classifies engineering design methods used in the processes of **Figure 2** into two major complementary groups: the creative, and the rational ones. The former are characterised by their ability to stimulate thought processes, removing mental blockages and widening areas of search for solutions to the design problem. The latter on the other hand, systematically examine different issues at each stage of the processes in Section 2.2, also eventually solving the same problem. It is reported that some creative people detest the latter approaches because of their apparent prescriptive nature. Many others however, find the rational approaches most helpful, even complementary to the creative ones. **Tables 1** and **2** summarise methods in these two groups of approaches.

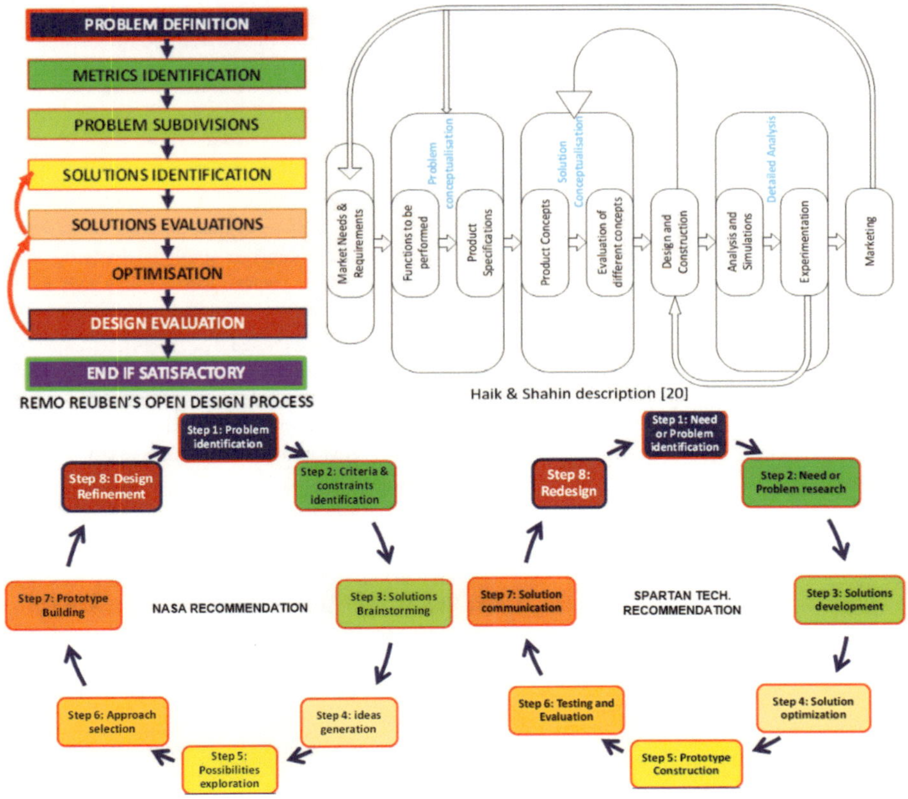

Figure 2.
Four examples of formal MED processes in industry.

Method	Comments
Brain storming	Best in a group, where each individual can make as many suggestions as quickly come to mind, irrespective of their apparent merits/demerits.
Synectics	Requires drawing parallels between quite unrelated events or things, preferably by people in a group discussion, to open up brain cells interconnection channels that can more easily lead to a solution.
Search space enlargement	Redefines the spectrum in which solution is sought. This can be assisted by: questioning the basis of the problem; random actions and making parallels of their effects to the problem on hand; dialectical reasoning, etc.
Spark of moment	Needs individual to have been thinking about the problem for some time—as in solution of plastic-latex jointing in Section 3.1 below.

Table 1.
Summary of creative engineering design methods.

Process	Methods	Comments
Identifying market needs	Market research and analysis	Build product objectives tree as ordered sets of targets to be achieved, in order to satisfy market needs in the prevailing circumstances.
Formulating product functions	Function analysis	Looks at the objectives tree, then determines overall function of the product to meet the objectives. The function is further decomposed into sub functions and then, possible mechanical components are identified to do, and, to integrate these sub functions. Limits on what can be done are imposed in form of a system boundary.
Specifying product attributes	Performance specification	Determined from functions, independent of possible solutions—considering that different products, types and features could provide the same functions. Pahl and Beitz [27] checklist can be used to define attributes that should preferably be quantified in range form, to give a specification.
Synthesizing alternative concepts	Brain storming	As in creative engineering methods.
	Morphological charting	Components for each sub-function in function analysis are tabulated and then, different combinations of these tried, to give different products having the same overall functionality.
Evaluating alternative solutions	Weighted objectives	Each feasible concept is considered for its relative position on each objective. Then summation of weighted scores guides selection. The problem however, is that ordinal scaling can result. This must be changed to interval value scaling.
	Pugh's evaluation matrix [28]	A benchmark concept is chosen. The others are compared with it in turn for each objective on a −1, 0, +1 scale. Totals for each objective, are multiplied by a weighting factor, and then sums of scores for each concept, computed. The bench mark concept scores zero while the best one is that, scoring highest.
Detail design and construction	Drawings	This step involves sketches of layouts for different concepts; assembly and detailed drawings of components of the selected concept.
	Prototyping	May exist in four forms [20]: Mock-ups, Models, Prototypes, and Virtual—CAD generated systems. Experimentation is done on the first three while simulations are done on virtual ones.
Improving the optimized solution	Value engineering	Re-examines the selected concept with intent of either reducing delivery cost—without losing functionality, or increasing value and utility to the customer, or both.

Table 2.
Summary of rational engineering design methods.

3. Case studies

We will now illustrate two cases of using some of the above approaches in an academic—rather than—an industrial environment. The first case is by the author himself. It exemplifies the creative design approach, and addresses an issue in solar energy engineering, of maximising useful energy yields from a flat-surfaced solar energy harnessing device. The second case shows a rational design approach, as taught to students in attempt to change MED from an analysis driven course, to a synthesis driven one. It builds on student knowledge gained from designing and constructing a multispeed fluid mixing vessel. The student designs a system for essential oils extraction from African herbs.

3.1 Case 1: invention of a new hydro mechanism for sun tracking

A new hydro-mechanism for interconverting linear and rotary motion was invented—and is described in a South African patent by the Cape Peninsula University of Technology (CPUT) [29]. The primary motive of the invention was to create a mechanism that would be deployed in a novel single axis sun tracking device that relied on mechanical energy to turn a domestic home solar energy collecting surface during the day, and return it to a morning position any time before daybreak. **Figure 3** shows the mechanism being used in conjunction with a photovoltaic panel.

3.1.1 Design processes

The approach used was a slight modification of the Remo Reuben open process in **Figure 2**. There was branching at the stage of evaluating alternative concepts, which led to other, very different products altogether—discussed in Refs. [30, 31].

Figure 3.
The hydro mechanism driving a sun tracking PV panel (watch online video at: https://www.youtube.com/watch?v=79CKBxt_h-I).

3.1.2 Identification of need and formulation of engineering problems

A need for a new single axis sun tracking device, suitable for sub-Saharan Africa conditions of bi-hemispherical location, low credit and disposable incomes, and an inadequate technical skill base had been established in Ref. [32].

The product design problem and its sub problems were defined as: "Design a single axis sun tracking mechanism and its coupling means to a domestic home flat solar collector, so that the latter will be able to receive more energy from the sun, and therefore through the appropriate conversion process, yield more output than when in a fixed orientation."

The sub problems, imposed by constraints discovered during identification of the 'Need' were:

- What would be the source(s) of energy in the mechanism?
- What motion transformations would the mechanism have to effect—and by which machine elements?
- How would the motion transformation be controlled, and how much energy would be required for both transformation and control?
- Which materials and manufacturing/assembly methods would be used to make and install the mechanism?
- What operational and maintenance tasks would be expected of the owner/user?

3.1.3 Solutions: evolution of the gravity driven hydro-mechanical tracker

Many solutions were investigated. Some were tried up to manufacture stage, and then discarded. Here, only significant ones are described in chronological order up to the prototype milestone. The reasons for discarding or modifying them are given.

3.1.3.1 The initial concept: solar-thermal hydraulic (STH) system with mechanical clock control of valves

A STH powered, spring controlled system was envisaged as in **Figure 4**. A hydraulic head ***hmax*** was to be provided by a liquid in shaded tank '**A**', connected to spring bank ***k*** through a one way valve **V1** and a cylinder-piston assembly. The piston rod would be a cylindrical rack, which energises the compression spring at the other end. The rack would drive a gear train to which, the frame holding the solar collector would have been coupled. At the end of the cylinder, was to be a one way valve **V2,** leading to an un-shaded small tank '**B**'. In late evening, with **V2** closed, and the piston at extreme western position, the panel would be facing west. **V1** would be actuated to open by the closure of **V2**. Liquid would flow into the cylinder, and compress the spring bank to the maximum (at which point a head ***hmin*** would act on the piston) while reorienting the panel to face eastward for the next day. In the morning, valve **V2** would be actuated by a clock signal to open. Valve **V1** would simultaneously close. During the day, the **V1-V2** link would be automatically deactivated to enable independent operation of **V2** without opening **V1**. Then, a clock mechanism would intermittently open **V2** allowing a precise liquid weight to flow into un-shaded tank '**B**' and then close for a definite period. This would relieve the spring bank a distance ***x***, and in turn rotate the solar collector a definite angle westward—hence tracking the sun in this direction.

In due course, the liquid in tank '**B**' would vaporise and condense in shaded tank '**A**' so that the hydraulic head in '**A**' gradually rose during the day in preparation for a night

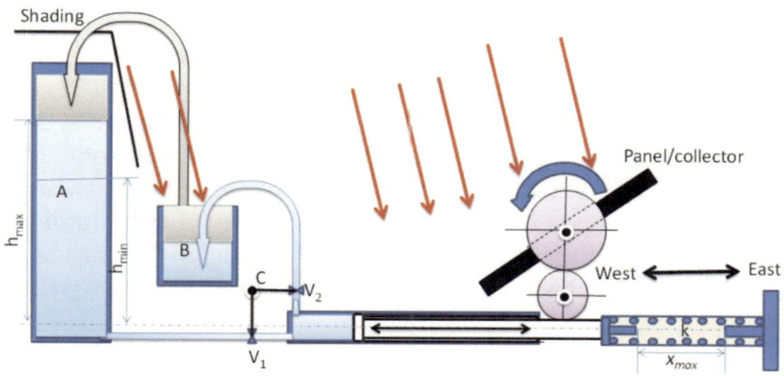

Figure 4.
Concept number 1: solar-thermal-hydraulic (STH) system.

return action. The panel tracking axis could be rotated about the piston-cylinder axis for correct installation at different latitudes by graduated scales in a plane perpendicular to the figure.

This concept required use of a low boiling point, low enthalpy of evaporation but high density liquid. The low enthalpy would give sufficient daily solar assisted evaporation rates while the high density would enable storage of enough mechanical energy to compress the spring and turn a collector whose centre of gravity would most likely be offset from the axis of rotation. Such a liquid was actually identified among the refrigerants (R140a) but it was expensive in Cape Town. Being a chloro-hydro-carbon (CH_3-CCl_3), it was banned in some African countries. Although there were other issues, this alone was sufficient to disqualify the concept. However, many of its elements were carried to the next concept.

3.1.3.2 Concept 2: solar-thermal hydraulics (STH): water replaces R140a

After discarding the concept of using a chloro-hydrocarbon, water was considered. The immediate problem however, was that it was less dense and had a much lower vapour pressure at the envisaged working temperatures. Most importantly, its enthalpy of evaporation was an order of magnitude higher than that of R140a. These limitations were to be overcome in a series of solutions—still using STH principles (i.e., evaporate the liquid, raise it to some height and condense it there to provide a head that will reset the mechanism at the end of the day). A summary of the salient 'solutions' up to the time the STH system was discarded is given below.

- Evacuation of the system so that the boiling point could be lowered significantly to say, below 60°C. This was in attempt to raise the vapour pressure at a working temperature of between 30 and 40°C in the evaporator tank '**B**' of **Figure 4**.

- A redesign of the evaporator '**B**' to a flat plate solar collector, possibly able to evaporate at least 5 kg of water on a 'typical' Cape Town winter day. This was to be supplemented by provision of hybrid heating using say, biomass below the evaporator—in case the day was very cloudy.

- A redesign of the condenser tank '**A**' and its condensation means to use evaporative cooling so that a smaller unit could be used. This was followed by elevating the tank to an appropriate height and by design of a vapour evacuation and piping system from the evaporator.

- A redesign of the mechanical linear to rotary motion inter-conversion system. A light weight semi cylindrical rack was to be in rectilinear motion, atop a rigid stem. The stem was to be attached to the spring loaded piston. The rack would then drive a fixed axis spur gear, mounted on the solar collector's axis of rotation (**Figure 5**). For locations say in the southern hemisphere, one side of the rack would be used. The other half would be used in the northern hemisphere, where the orientation of the axis and relative position of evaporator would have to be switched to still enable east to west day tracking. In this way, no internal readjustments would be necessary, if the device was moved across the equator.

Figure 5 illustrates the mechanism at this stage. The mechanical valves have also been replaced with solenoid valves by now.

STH systems had been attractive mainly because they looked novel and relied entirely on 'free' solar energy for their operation most of the time. They had a simple backup plan of burning biomass in case of cloudy days. The evaporator, the vapour evacuation system, the cylinder-piston-spring assembly were designed and constructed. A 200 mm × 200 mm × 100 mm aluminium block for manufacture of the semi cylindrical rack was also purchased. Meanwhile, a separate experiment to verify findings of a theoretical analysis on water evaporation rate yields in an evacuated collector gave 'unwanted' results. Whereas water seemed to evaporate fast enough at the low pressures, most of the vapour re-condensed on the collector glazing and in the evacuation piping before reaching the condenser. It was clear that a more elaborate evacuation system would have to be used if STH were to progress further.

A second 'unwanted' result came from the workshop. Machining of the cylindrical rack in the CNC workshop encountered problems when the purchased block was being resized for actual machining (it had not been exactly 200 mm × 200 mm × 100 mm). These problems forced a re-examination of the 'needs' of Section 3.1.2. It became apparent that the manufacturing problems being encountered, together with the possibility of vacuum leaks in the field would make the product not only 'too expensive', but would also affect its reliability. Moreover, as seen in **Figure 5**, the mechanism would be bulky, and perhaps less marketable than substitutes which could come on the scene later. Thus, STH on this product was discarded. Use was however to be made of almost all components and learnings from it in this and other off shoot products.

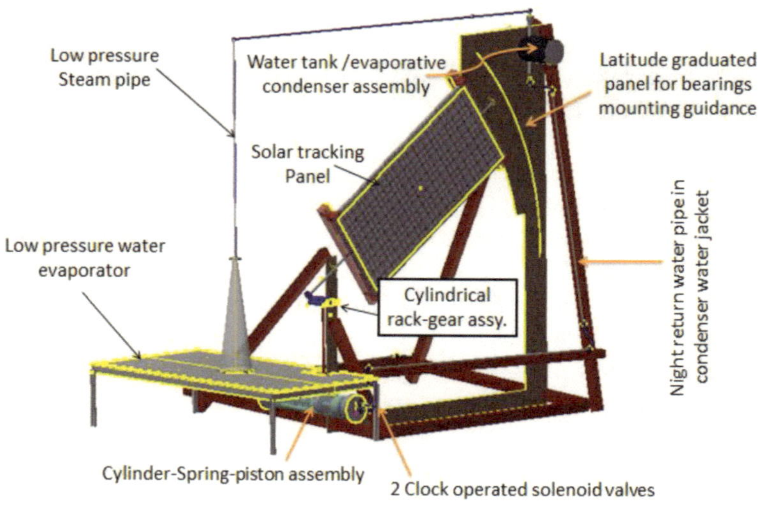

Figure 5.
The last of the STH concepts.

3.1.3.3 Other concepts: genesis of the hydro-mechanism

Although STH was now out of the question, the idea of a hydraulic head provided by an oversized 100 L condenser in **Figure 5** still remained attractive. The condenser had intentionally been oversized to provide sufficient heat transfer area, and also to hold reserve water in case of bad weather and inability to light a fire under the evaporator for whatever reason. The piston-cylinder assembly had been designed to discharge about 5 L a day—which would have easily been evaporated by energy incident on a 1.8 m × 1.2 m collection surface. It was therefore reasoned that with 100 L initially filled into the condenser tank (by whatever means), there could be a 20 day pumping head capacity to reset the mechanism at night. The 50 L tank of **Figure 4** was also revisited to hold daily discharges from the mechanism. This would therefore hold slightly more than a week's discharge (as it could not be filled to capacity). This, at last seemed to settle the hydraulics part—if only the cylindrical rack could be made. It was not made.

3.1.3.4 The double rack replaces the cylindrical rack: a Hooke joint appears; the cylinder orientation changes

Because of manufacturing difficulties mentioned above, the aluminium rack design was reconsidered. Moreover, in absence of the evaporator, the stem sticking out of part of the cylinder looked neither a safe nor an aesthetically 'correct' design. Therefore, it was decided to use an ordinary straight rack-gear set completely housed within the cylinder. The rack would now be part of the piston rod, while the gear shaft axis would be fixed. The gear shaft would protrude slightly out of the cylinder to connect to the collector shaft. Minding about the 'Needs' in Section 3.1.2 on deployment anywhere in sub-Saharan Africa and the now user-inaccessible gearing, the gear shaft was to be standardized as horizontal, normal to the cylinder axis. Variable slope collector shaft axes for different locations were to be joined to this horizontal shaft with a Hooke coupling. Bi-hemispherical installation was to be facilitated by a double rack-gear set such as shown in **Figure 6**.

This selection of elements would affect the geometry of the mechanism-collector connection of **Figure 5**. Either the collector would have to be lowered, or a horizontally oriented spring-piston-cylinder assembly would need to be raised. The former was considered impractical because a collector-ground clearance must be maintained

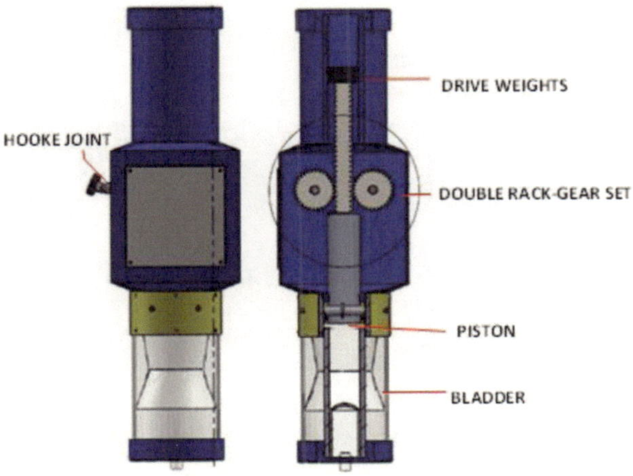

Figure 6.
Reorientation of the hydro-mechanism.

during all phases of rotation of the collector. The latter was considered aesthetically unsound and required more ground space to effect. Therefore, the orientation of the cylinder was now changed to vertical as in **Figure 6**.

3.1.3.5 A pump is introduced: the raised tank is temporarily ignored

At this stage, it was supposed that the 3.5+ m high tank could be installed either on a roof or on a stand provided with properly constructed ladders for filling and inspections. In rural Africa, once about 20 days, the owner or her/his agent would have to climb up and refill it. At the university however, experimentation in the project required a safer and smarter way of filling the tank. The 'Needs' constraints of Section 3.1.2 specified a 'negligible' energy consumption by operation of the mechanism. Since tank filling would be occasional, a small 12 V DC 4 m peak head pump was acquired. Then for experiments, only one tank ('B' of **Figure 6**) would be necessary. The pump would be used to transfer water from this tank to the mechanism. In addition to being safe, this would conserve water since it could be recycled on daily basis.

3.1.3.6 The piston-cylinder part fails: bellows appear: and also fail

The mechanism was now ready for prototyping. The machine elements and components were assembled. First attempts to run it were made in March 2015. Water leaked past the piston. To properly seal the leakage, it would be necessary to reduce clearances. A new piston with an elastomer O-ring was made. Sealing was achieved but friction was excessive. A lot of pump energy went into overcoming this friction. Then in one re assembly, a forceful push onto the piston burst the cylinder. **Figure 7** shows the broken piece. It was now evident that the engineering necessary to produce an efficient and reliable piston-cylinder assembly would easily 'violate' the 'cost effectiveness' constraint in Section 3.1.2, and probably consume more than 'negligible' energy in operation. The assembly had to be redesigned.

Friction between the cylinder and the piston was the main problem in the assembly. It was therefore decided to eliminate direct contact between the piston and the cylinder. Bellows were introduced. One end of the bellow was fixed to the lower base of the cylinder while the other was fixed to a smaller diameter piston. It was supposed that water would progressively fill the bellow segments starting with the lowest, and in so doing, gradually lift the piston-spring ensemble without any significant friction with the cylinder. The first bellow tried was made from

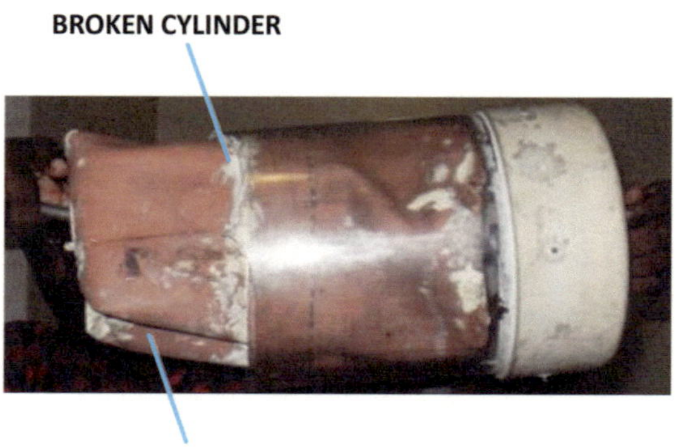

Figure 7.
Example of failure during the project.

a 150 mm diameter heli-steel PVC hose. It was readily available and 'reasonably' priced at just below the equivalent of US$ 5.00 a meter length. It however, failed on trial. When water filled the first segments, they expanded radially before attempting to lift the piston. Even after lift-off, the expansion continued until the steel was beginning to tear out. Attempts were made on using a thicker and stronger rubber bellow made by a local rubber products moulder. It also failed. It was clear that for the bellows to be of use in this project, they would have to be restrained radially—which in the circumstances, was not feasible. They were abandoned, but lessons on need for radial restraint were to serve a breakthrough purpose soon.

3.1.3.7 The bladder is born: the spring works but it is retired; gravity takes over

Bellows failed because they were not restrained radially. Even if they had not failed, it was difficult to tell what would happen to the joints at the base in rural Africa over a prolonged period of intermittent pressurization. It was therefore decided to contain the mechanism water in a flexible liquid sac or bladder that would be completely restrained and protected by a much stiffer, though flexible covering. The active part of the bladder would be an inverted cone frustum grown on a lower normal frustum which in turn, would have grown on a cylindrical portion matching the internal surface and base of the mechanism cylinder. The cylindrical and lower cone frustum would always be with water. Pumping would only affect the upper frustum which would be closed by a permanently joined and sealed piston. The piston would carry a small bleed pipe as shown in **Figure 8**. The primary purpose of this pipe would be to help expel air from the system on first fill.

The bladder and its protective covers were constructed and assembled in the mechanism. The system was then test run. At long last, it was able to reach its design peak compression on 2nd July 2015. But the time to reach maximum displacement was in excess of 2 min. Also, towards that endpoint, the 10.8 W pump was drawing maximum current. The top mechanism end cap was removed so that the mechanism could be filled with water against the spring and piston weights only. It took about 20 s to reach the top dead centre position. On opening valve **V2** of **Figure 4** to simulate a day time operation (i.e., rotate the collector East to West), the collector turned the full 180°, but there was difficulty in traversing the mid position. This meant the spring weight (30.2 N) was barely enough to drive the mechanism. However, it had been established that spring force (***kx***) was not necessary to run the mechanism—and therefore the spring could be discarded to give way to gravity weights. A gravity driven, bladder-flow regulated hydro mechanism for sun tracking had just been invented.

3.2 Case 2: design of a herbal oil extractor for small scale industries

This case study is about a project which started off as one of the many group projects in normal class time, intended to overcome the familiar

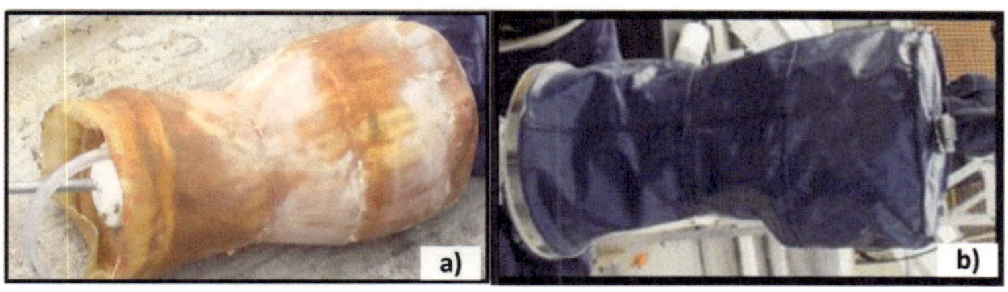

Figure 8.
The prototype bladder: (a) un-protected and (b) when protected.

Figure 9.
A home cottage industry 'Two speed' fluids mixer—as designed by MED students left—the assembled unit. Right—the counter-rotating slow speed mechanism (Watch online video at: https://www.youtube.com/watch?v=VymLqK9DDeo).

'analysis-synthesis' barrier in undergraduate MED. A group of six students had initially been tasked and guided to design and construct a variable temperature and viscosity fluids mixer for a home-cottage cosmetics factory within a period of 6 weeks. The mixer is shown in **Figure 9**. After the project, one of the students was involved in a serious road accident which disabled him, and prevented him from doing the normal pre graduation industrial attachment. To enable him graduate however, he was assigned a new individual design project under supervision of the author at the university. He was to use his experience in the class project, to design (not construct) a herbal oil extractor, again for a home cottage factory. Below is a summary of his design approach.

3.2.1 Identification of needs and formulation of engineering problem(s)

A machine needed to be designed for use in extraction of essential oils from African herbs. A full design with drawings (mechanical, electrical and hydraulic) was to be completed so that students could manufacture and test the machine.

3.2.1.1 Requirements and constraints

- Handles 200 L with 30% spare capacity.
- Filtered liquid product must be extracted separately from the spent herbs.
- The extraction temperature must be between 70 and 80°C.
- Operates on domestic single phase 220–240 V AC power supply.
- Feed herbs are received cut into pieces smaller than 10 mm in length.
- Professional and pleasing appearance
- Students must be able to manufacture all custom designed parts in the CPUT mechanical engineering workshop.

3.2.2 Concepts development and selection

The student considered five concepts as shown in **Figure 10**. He settled for concept number 3 on account of ease of manufacture, minimal heating element corrosion risks, and maximum heat transfer area, thereby reducing heating time.

3.2.3 Detailed design and specifications

Having selected the mode of heating for the herbs, the student laid out the design as in **Figure 11**. Then he did a detailed analysis of the chosen concept in virtual form to give specifications in **Table 3**, followed by detailed engineering drawings of each machine element in the system.

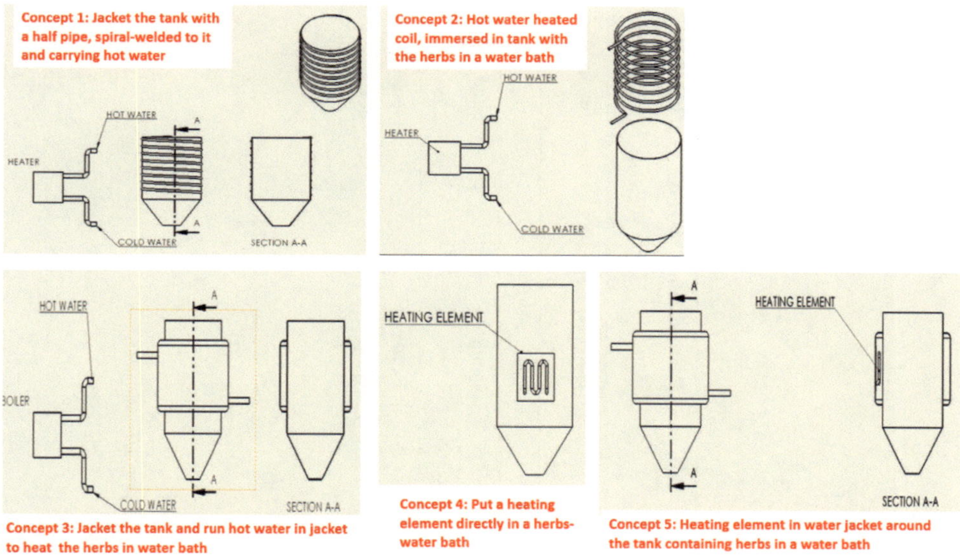

Figure 10.
Herbal oil extraction: different heating concepts [33].

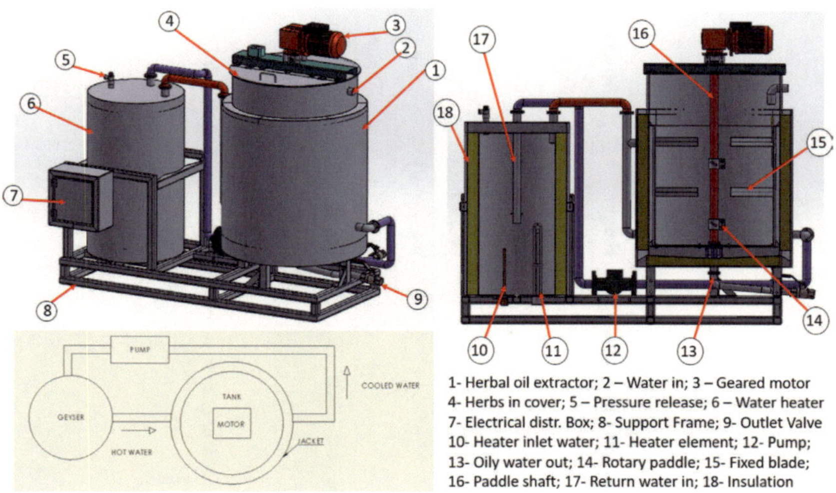

Figure 11.
System layout for a small scale herbal oil extractor.

Overall dimensions (mm)	Height	Length	Width	Power at 240 V AC (W)	Heater	Motor	Pump	Total
	1478	**1660**	**930**		**4000**	**180**	**260**	**4400**
Motor Specs	Type	Manufacturer	Code	Power (W)	Torque (Nm)	Speed (rpm)	Output speed (rpm)	
	Worm geared	SEF	SF37DRK7154	180	70	1450	15	
Heater	Power (W)	Manufacturer	Code	Pump	Head (m)	Power (W)	Max flow (L/s)	
	4000	Thermon	UHRBC14K00350N1		2.75	260	2	
Temperatures (°C)	Heater max.	Vessel max	Materials	Heater	Vessel	Insulation	Frame	
	90	70		SS316	SS316	Thermalite Glass wool	MS	

Table 3.
Key specifications of components for a 200 L herbal oil extractor for a home cottage factory.

4. Summary and conclusion

In this chapter, we have described and illustrated MED from both a classroom perspective and from an industrial one. In undergraduate MED, emphasis is on ability to analyse existing systems. The student is taught MED on a machine element by element basis—and most contact hours are spent that way. There is little room and time to integrate the elements in one worked example or problem. Moreover, those elements from other areas of mechanical engineering, such as in Thermo-Fluids, are normally assumed to be well covered in those subjects. They are rarely given consideration in normal MED class rooms. This is not to mention the even more critical considerations of non-science related issues which, in the first place, are often, the source of problems to be solved by engineering. It has been argued in this chapter that this treatment acclimatises the student to always be expectant of readymade systems to analyse. Even then, understanding how the systems came into being, as answers to specific human needs, can be problematic. This is a disservice to the student and to industry, because the main purpose of MED is supposed to be synthesis of mechanical systems, speaking to the needs of society.

Industry on the other hand, has no choice but to face the design challenge from a problem solution perspective. All areas of knowledge, be they from science, art, or even heuristic and intuition, are brought to bear on the problem. Market, economic, political, legal, social, aesthetic and ordinary engineering constraints are imposed on an inexistent system that is supposed to be created and made. Two largely complementary approaches of doing so were reviewed: the creative, and the rational. It was seen that the rational approach formalises the design process and tends to take care of more constraints a design may be encountering. It is thus advisable, even of creative designers to embrace it. In engineering classrooms, it is without a doubt, the recommended approach.

Two design examples were described in a university setting environment. One was primarily of creative nature, leading to an invention over a long period of time. However, whether consciously or otherwise, it was still tempered with some formality in form of a structured approach between different design stages. The main advantage of this approach seemed to be the generation of other offshoot products arising from apparent failures within the creative process. In essence, therefore, creativity can lead to many other originally unintended, but useful products. The second example was focused on the rational approach—as taught to the author's students. A physically handicapped student was able to demonstrate that he had learned the methodology by designing a product quite related to what he had learnt—and participated in building in class, while still physically fit. Importantly, he demonstrated good understanding of the integrative nature of MED, calling on subject content from diverse areas like Fluid Mechanics, Heat transfer, Electrical Technology, Economics, etc. Moreover, the problem to be solved required him to appreciate compositions of some naturally occurring plants and means of getting useful extracts from them. Such extensive exposure is not normal in MED as commonly taught/learnt.

To conclude the chapter, it could be said that—although the traditional approach of handling MED is helpful in so far as it breaks up the subject matter into smaller, easier to learn, topics, it makes it more difficult for students, and possibly academics, to apply that knowledge to solve real life engineering design problems. This author recommends a mixed approach whereby, early on in the study of MED, the current topic-based system is used but a formalised needs-driven design approach is gradually introduced until it becomes the dominant approach by the time the student is finishing her/his MED subjects. **Table 4** summarises the salient differences in approach.

It is to be understood here, that we are not talking about the final year design project, typical in many engineering schools and faculties. No—it is the timetabled MED we are referring to. The new approach not only works, but it produces tangible results as demonstrated in this chapter. It should therefore, as much as possible, be adopted.

	Traditional	Proposed
Content	MED as extension of Engineering Science Low involvement of issues in Humanities	MED as new body of knowledge, using GEPs*, Design codes, catalogues, standards, etc. Elements from Humanities primarily drive most of MED and even, regulate it
Modelling	Simplified mathematical models and physical science phenomena	Mathematical and non-math models to deal with science and non-science constraints in design optimization
Learning	Study of independent machine elements	Less time on independent elements, but more on their interdependence in an assembly
Design process	Element to assembly design approach (down-up)	Assembly to element design approach (top-down)
Emphasis	Engineering and scientific novelty	Problem solution as simply, safely and economically as possible
Outputs	Engineering drawings	Working models and/or prototypes

**Good engineering practice principles*

Table 4.
Summary of differences between the current and proposed MED teaching approaches.

Acknowledgements

The work in case No. 1 was funded by Cape Peninsula University of Technology Research Fund, through research account RK23.

Conflict of interest

The author's interest in this, and other related work, is driven by an insatiable desire to make students realise that by their last year of undergraduate study, they can already have an inner ability to start contributing to make their societies live better now, and not wait for tomorrow.

Thanks

The author thanks his student, Riel Haupt, for his drive and courage even after the almost fatal accident. Special thanks go to Riel's parents for supporting him.

Author details

Kant Kanyarusoke
Cape Peninsula University of Technology, Cape Town, South Africa

*Address all correspondence to: kanyarusokek@cput.ac.za

References

[1] Joyce T, Evans I, Pallan W, Hopkins C. A hands-on project-based mechanical engineering design module focusing on sustainability. Engineering Education. 2013;**8**(1):65-80. DOI: 1011120/ended.2013-00008

[2] Kanyarusoke KE. Spreadsheets in mechanical engineering design: Getting students to learn and to do the 'real stuff'. In: Proceedings of 4th Biennial SASEE Conference; Cape Town South Africa. 2017. pp. 399-402. ISBN: 978-0-620-57123-4

[3] Wickert J, Lewis K. An Introduction to Mechanical Engineering. 4th ed. Stamford, CT: Cengage; 2017. 432 p. ISBN: 9781305635753

[4] Cohen K, Katz R. Teaching mechanical design practice in academia. Procedia CIRP. 2015;**36**:177-181

[5] Katz R. Integrating analysis and design in mechanical engineering education. Procedia CIRP. 2015;**36**:23-28

[6] University of Cape Town. Faculty of Engineering and the built environment (Undergraduate) [Internet]. 2017. Available from: www.students.uct.ac.za/usr/apply/handbooks/2017/EBE_UG_2017.pdf [Accessed: November 9, 2018]

[7] Juvinall RC, Marshek KM. Fundamentals of Machine Component Design. 6th ed. Hoboken, NJ: Wiley; 2018. 720 p. ISBN: 1119382904, 978119382904

[8] Budynas RG, Nisbett JK. Shigley's Mechanical Engineering Design. 10th ed. New York, NY: McGraw Hill; 2015. 1082 p. ISBN: 9789814595285

[9] Mott R, Varvek EM, Wang J. Machine Elements in Mechanical Design (What is New in Trades & Technology). 6th ed. New York, NY: Pearson; 2018. 880 p. ISBN: 100134441184; 139780134441184

[10] Madisen DA, Madisen DP. Engineering Drawing and Design. 6th ed. Mason, OH, Cengage; 2016. 1680 p. ISBN 10: 1305659724; ISBN 13: 97881305659728

[11] Hibbeler RC. Engineering Mechanics Dynamics. 14th ed. Hoboken, NJ: Pearson; 2016. 791 p. ISBN 10: 0133915387, ISBN 13: 9780133915389

[12] Moran MJ, Shapiro HN, Boettner DD, Bailey MB. Fundamentals of Engineering Thermodynamics [Internet]. 8th ed. 2014. Available from: https://www.academia.edu/21431299/FUNDAMENTALS_OF_ENGINEERING_THERMODYNAMICS_Eighth_Edition?auto=download [Accessed: November 9, 2018]

[13] Hughes E. Electrical and Electronic Technology. 12th ed. Harlow, England: Pearson; 2016. 1008 p. ISBN 10: 1292093048; ISBN 13: 9781292093048

[14] Simbo C. Defining the term basic education in the South African constitution: An international law approach. Law Democracy & Development. 2012;**16**:162-184

[15] Mukwinja VM. Rethinking education in Botswana: A need to overhaul the Botswana education system. Journal of International Education Research. 2017;**13**(2):45-58

[16] Jwan JO. Republic of Kenya Basic Education Curriculum Framework. [Internet]. 2017. Available from: www.kenyaexamcenter.com/pluginfile.php/373/mod_resource/content/3/Curriculum%2OFramework$202017.pdf [Accessed: November 14, 2018]

[17] Mathew IA. Provision of secondary education in Nigeria: Challenges and way forward. Journal of African Studies and Development. 2013;**5**(1):1-9

[18] GOU (Government of Uganda). Education Sector Profile. [Internet] 2018. Available from: www.ugandainvest.go.ug/uia/images/Download_Centre/SECTOR_PROFILE/Education_Sector.pdf [Accessed: November 18, 2018]

[19] Pindula—New Education Curriculum Zimbabwe [Internet]. 2018. Available from: www.pindula.zw/New_Education_Curriculum/_Zimbabwe_Centre/SECTOR_PROFILE/Education_Sector.pdf [Accessed: November 18, 2018]

[20] Haik Y, Shahin T. Engineering Design Process. Stamford, CT: Cengage; 2011. 287 p. ISBN 10: 0495668168; ISBN 13: 9780495668169

[21] Cross N. Engineering Design Methods Strategies for Product Design. Chichester, England: Wiley; 2008. 230 p. ISBN 10: 9780470519264, ISBN 13: 9780470519264

[22] Ullman DG. The Mechanical Design Process. 6th ed. New York, NY: McGraw Hill; 2017. 480 p. ISBN 10: 97809999358804, ISBN 13: 9780999357804

[23] Ertas A, Jones JC. The Engineering Design Process. 2nd ed. New York, NY: Wiley; 1996. 628 p. ISBN 10: 0471136999, ISBN 13: 9780471136996

[24] Spartanteched. Engineering Design Process—Remember the Order [Internet]. 2012. Available from: http://spartanteched.wikispaces.com/Engineering+Design+Process+-+Remember+the+order [Accessed: August 4, 2016]

[25] NASA. Engineering Design Process [Internet]. 2015. Available from: http://www.nasa.gov/audience/foreducators/plantgrowth/reference/Eng_Design_5-12_prt.html [Accessed: August 4, 2016]

[26] Reuben R. Remo Reuben's Engineering Design Portfolio [Internet]. 2013. Available from: HTTPS://REMOENGG.FILES.WORDPRESS.COM/2013/04/PROCESS.PNG [Accessed: August 4, 2016]

[27] Pahl G, Beitz W, Feldhusen J, Grote K-H. Engineering Design: A Systematic Approach. 3rd ed (Translated by Wallace K, Blessing L.). London, England: Springer; 2007. 617 p. ISBN 10: 1846283183

[28] Burge S. The Systems Engineering Tool Box. [Internet]. 2012. Available from: https://www.burgehugheswalsh.co.uk/uploaded/1/documents/pugh-matrix-v1.1.pdf [Accessed: November 8, 2018]

[29] Kanyarusoke KE, Gryzagoridis J. A hydraulic drive mechanism, South African provisional patent No.P72557ZP00 for Cape Peninsula University of Technology. South Africa Patent Journal. 2016;**49**(1):12 (part II of II)

[30] Kanyarusoke K. Solar water purifiers for small rural African homesteads: Evaluation of alternative designs. In: IOP Conference Series: Earth and Environmental Science. 188 012004. 2018

[31] Runganga T, Kanyarusoke KE. Design, construction and performance testing of solar crop dryers for rural areas. In: Proceedings of 26th ICUE Conference; Cape Town South Africa. 2018. pp. 42-49. ISBN: 9780994675958

[32] Kanyarusoke KE, Gryzagoridis J, Oliver G. Are solar tracking technologies feasible for domestic applications in rural tropical Africa? Journal of Energy in Southern Africa. 2015;**26**(1):86-95. ISSN: 1021-447X

[33] Haupt R. Design of a Herbal Oil Extractor [ND project report]. Cape Town, South Africa, Cape Peninsula University of Technology; 2018

Section 2

Innovation Methods in Naval Engineering

Chapter 6

Prologue: Exploring the Potential of the Sea

Sérgio Lousada and Rafael Camacho

1. Naval engineering

Naval engineering is the branch of engineering that has as main activity the exploration of the potential of the sea. Although specialized, naval engineering is quite eclectic since it addresses the main aspects of other engineering modalities, directly or indirectly. With the development of oil exploration and production in the ocean, the naval engineer's work extended to ocean engineering.

The naval engineer is engaged in all stages of life of vehicles and platforms for maritime transport, the exploitation of marine resources and recreational activities, from the concept and design phase, to the construction phase and its quality control, through inspection, maintenance and repair. Also included are maritime and port operations planning and management tasks. The professional must have the systemic and comprehensive vision necessary for the design of large engineering systems.

Appealing to the wide-ranging aspect of naval engineering, the engineer also deals with the design, construction, repair and logistics of inland waterways used for inland transportation.

2. Fields of intervention

The fields of intervention of the Naval Engineer are quite broad, concerning design, construction, modification, repair, maintenance and dismantling of ships, vessels, other maritime vehicles and floating structures [1].

In the project phase, the Naval Engineer has overall technical responsibility, including safety and pollution prevention, as well as aspects related to the nautical and operational qualities of the ship, as well as the ergonomic and automation aspects. In these areas, the Naval Engineer performs calculations in the various technical areas of his specialty, acts as systems integrator on the ship and as coordinator of multidisciplinary teams made up of engineers and other professionals who deal with the numerous sectoral aspects [1].

In the industrial activity of construction, modification, repair and dismantling of ships, the Naval Engineer acts either as a production or project manager or as a specialist in the planning of production and in the definition of constructive strategies and operational solutions. In this context, it also carries out docking and launching calculations, defines and directs the stability test, performs its calculations and defines and directs the tests to the pier and sea [1].

IntechOpen

Other fields of intervention of the Naval Engineer include the following [1]:

- Technical, commercial and maintenance management;
- Management and planning of maritime operations and the maritime-port interface;
- Development and application (approval, survey and inspection) of rules and regulations;
- Consulting and technical advice, as well as expert activity;
- Marketing and technical support of marine systems and equipment;
- Teaching and research.

3. Technical areas

The technical areas of Naval Engineering are the different interdependent areas that allow to carry out the development of the ship project in its entirety, being the following [1]:

- Naval Architecture;
- Statics and Stability;
- Resistance and Propulsion;
- Structural Strength;
- Behaviour at Sea;
- Maneuverability;
- Vibration and Noise.

The Naval Architecture comprises the design of the ship, the development of the general arrangement and a first approximation of the geometric plane, considering both the design requirements and the classification rules and legal regulation applicable. The interior spaces of the ship are allocated to the various functions, contemplating functional requirements, fire and flood protection, escape routes, life-saving appliances, multiple ship systems, navigation and communications equipment and the signaling systems which must be integrated into a functional and coherent general arrangement [1].

In the Static and Stability area, the buoyancy and stability of the ship (intact and damaged) shall be assessed in accordance with the applicable classification rules and legal regulations in order to ensure a safety margin against tipping actions that the ship may encounter in its life: waves, wind, cargo or others [1].

The Strength and Propulsion area consists of the dimensioning of the propulsion system (main machines, reducers, shafts, propellers or other propulsion means) to achieve the design speed, considering the geometric plane, with the highest energy efficiency and minimizing emissions [1].

The Structural Strength area considers the dimensioning of the structure of the ship in all its details, including type of materials and scantlings, considering its general arrangement, in order to ensure the necessary resistance to stress due to waves, cargo or others, considering also the requirements of the applicable classification rules [1].

The Sea Behaviour area consists of studying the movements of the ship under the action of waves, regarding movement amplitudes, accelerations, propeller emersion, bow emersion, water boarding on the deck, in order to assess the operationality of the ship in sea [1].

The area of Maneuverability consists of the study of maneuverability and directional stability of the ship, mainly for maneuverability in confined waters, also considering applicable legal requirements [1].

The area of Vibration and Noise studies the vibration and noise levels in the various on-board spaces, as well as the radiated noise to air or water, determining the measures necessary for their minimization and maintenance within the limits imposed by classification rules or legal regulations [1].

4. Cultural inheritance

The Naval Engineers in Portugal are heirs of a long tradition in shipbuilding, which allowed the Maritime Expansion in the fifteenth and sixteenth centuries. The legacy of some of the men who studied and built ships in Portugal at that time have reached our days: Fernando de Oliveira, João Baptista Lavanha and Manuel Fernandes. The Portuguese Age of Discovery and later maintenance of the regular transport of passengers and goods and maritime safety between the multiple components of the "Portuguese Space" up to the twentieth century are very much in keeping with Portuguese Naval Engineering [1].

The Portuguese Economy of the Sea has shown, over time, great economic and social resilience, being able to grow above the average of the national economy during a period of great economic instability. If investment in the marine industries is strengthened, particularly in terms of external investment, the economy of the sea will be able to progress to higher levels of growth and economic development [2].

Compared with its terrestrial dimension, the maritime geographic dimension of Portugal is enormous. Its location in the North Atlantic, in the west of the European continent, as maritime neighbors of the American and the African continents, and with a history closely linked to the oceanic routes, particularly with the Asian continent, indicates that its geographical position has relevant strategic significance [2].

The recent "rediscovery" of the Economy of the Sea as a source of wealth and resources, mainly with the development of marine renewable energies and with the expansion of the continental shelf, which could potentially contain oil, natural gas and minerals, will certainly pose great challenges and offer new opportunities to Portuguese Naval Engineering in the twenty-first century [1].

Author details

Sérgio Lousada[1,2,3] and Rafael Camacho[1,4*]

1 Faculty of Exact Sciences and Engineering (FCEE), Department of Civil Engineering and Geology (DECG), University of Madeira (UMa), Funchal, Portugal

2 VALORIZA—Research Centre for Endogenous Resource Valorization, Portalegre, Portugal

3 Institute of Research on Territorial Governance and Inter-Organizational Cooperation, Dąbrowa Górnicza, Poland

4 Investimentos Habitacionais da Madeira, EPERAM, Funchal, Portugal

*Address all correspondence to: rafaelfreitascamacho@outlook.com

References

[1] Ordem dos Engenheiros. Colégios e Especialidades. Naval [Internet]. 2019. Available from: http://www.ordemengenheiros.pt/pt/a-ordem/colegios-e-especialidades/naval/ [Accessed: 15-02-2019]

[2] PwC Portugal. Publications. HELM—Portugal [Internet]. 2019. Available from: https://www.pwc.pt/en/publications/helm-portugal.html [Accessed: 15-02-2019]

Chapter 7

Vortex-Induced Vibration of a Marine Riser: Numerical Simulation and Mechanism Understanding

Xiangxi Han, Youhong Tang, Zhiqiang Feng, Zhanbin Meng, Ang Qiu, Wei Lin and Jiaming Wu

Abstract

Marine riser is a key equipment connecting a floating platform and a seabed wellhead. Vortex-induced vibration (VIV) is the main cause of the fatigue damage of the riser. The prediction of marine riser VIV is very difficult because of its strong non-linearity, instability and uncertainty. In recent years, many numerical models of VIV of marine riser have been developed to explore the mechanism of marine riser VIV, providing scientific theoretical basis and practical engineering methods for vibration control and engineering design of marine riser. Combined with the authors' own recent research, this chapter discusses the research progress on marine riser VIV in the ocean engineering, including phenomenon mechanism analysis and different numerical research methods.

Keywords: fluid-structure interaction, marine riser, numerical simulation, structural dynamics, vortex-induced vibration

1. Introduction

As a key part of the deep-sea oil and gas exploitation system, the marine riser known as the "life line of offshore oil" is an important transport component connecting the undersea oil field with the ocean platform, as shown in **Figure 1**. In the process of operation, the marine riser bears the impact, erosion and other damage of seawater, and it needs to be repaired or renewed frequently. Its service life is only 2–3 years. Marine riser related technology has become one of the main obstacles to human being's deep-sea development. As the key structure connecting floating platform and underwater production system, marine riser generally bears the axial tension. The lower end connects with the seabed wellhead through the universal joint, and the upper end is coordinated with the slip joint on the platform or the bottom of the ship to carry out drilling, liquid guiding, mud guiding and other work. It is a necessary equipment for the deep-sea oil and gas development and transportation. In deep-sea environments, currents passing through the riser may induce the rear edge of the riser releasing the vortex alternately, known as "Karman vortex street". The presence of "Karmen vortex street" causes the riser structure vibrating simultaneously in the cross-flow (CF) and the in-line (IL) directions, and the vibration of the

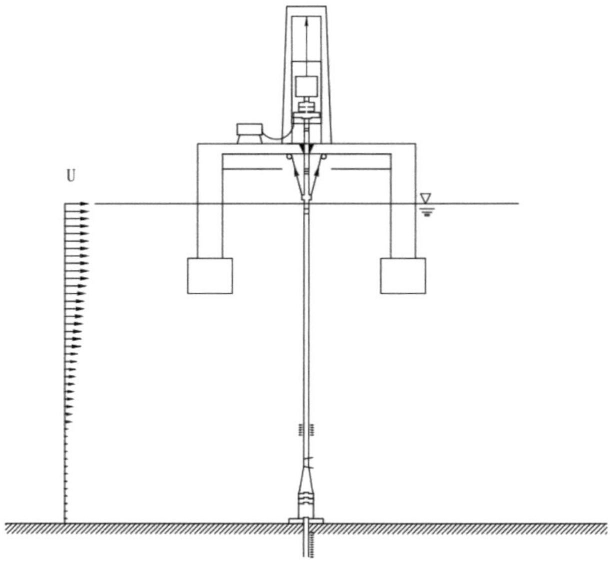

Figure 1.
Schematic diagram of marine riser operation.

riser will in turn change the vortex characteristics of the flow field. This phenomenon of fluid structure interaction is called "VIV". When the frequency of vortex release is close to the natural frequency of the riser structure, the phenomenon of "Lock-in" occurs. With the significant increase of the vibration amplitude of the riser, fatigue damage and even failure of the riser structure are often caused. In 2011, the "Gannet Alpha" drilling platform of shell oil company on the North Sea area of the United Kingdom suffered a sudden accident, resulting in the fatigue fracture of the oil pipeline connected with the platform. As a result, about 21.6 billion tons of crude oil leaked into the sea, causing serious pollution to the surrounding environment and causing major economic losses. However, the strong non-linearity, non-stationarity and uncertainty of the VIV of marine riser make it extremely difficult to predict the VIV of marine riser. How to accurately predict the VIV of marine riser and provide a more comprehensive theoretical basis for the design and daily maintenance of marine riser has been an ideal goal pursued by engineers and scientists.

For the study of VIV of marine riser, early scholars started from the simplest riser section, and now they have carried out a very rich experimental and numerical simulation study on two-dimensional (2D) rigid cylinder, including the study of flow around fixed cylinder, forced vibration and self-excited vibration of cylinder. The study of a rigid cylinder VIV (2D VIV) establishes many basic concepts of VIV, explores the law of rigid cylinder VIV, and lays a foundation for the construction of three-dimensional (3D) riser VIV algorithm model. However, the vibration of marine riser shows the characteristics of high order vibration and the vortex released by the rear edge of the riser has obvious 3D characteristics. Therefore, the research results based on the 2D VIV fail to provide complete technical support for the 3D marine riser. This prompted the researchers to study the VIV of 3D marine riser based on the research results of a rigid cylinder VIV. Due to the different emphases of the study on the VIV of a rigid cylinder and slender flexible riser, there are also many differences in the research methods and research hotspots. So far, there have been abundant researches on VIV, and many reviews have summarized the results of studies on the VIV of a rigid cylinder or a riser. For example, Williamson [1], Williamson and Govardhan [2, 3], Sarpkaya [4], Gabbai and Benaroya [5], Bearman [6], Wu et al. [7] have summarized the research results of a cylindrical structure or a riser VIV in recent decades. In this chapter, the numerical simulation study of marine riser VIV in recent decades are summarized, and the development process, advantages and disadvantages of different numerical research methods are analyzed and summarized. Finally, the research progress of current basic research and hot issues in engineering practice is discussed. The numerical analysis methods of marine riser are mainly divided into two categories: one is the empirical model method based on experimental data, the other is the numerical simulation method based on computational fluid dynamics (CFD) and computational structure dynamics (CSD).

2. Empirical model

The numerical prediction method based on empirical model is suitable for solving practical engineering problems, because the physical model is clear, and the calculation is simple. Empirical models mainly include wake oscillator model, modal superposition model, statistical model, etc. Among them, wake oscillator model and its improvement model are most widely used. Bishop and Hassan [8] and Hartlen and Currie [9] proposed the wake oscillator model and described the hydrodynamic force with the nonlinear Van de Pol equation. In the model, the hydrodynamic equation is coupled with the vibration equation which describes the transverse VIV of the structure. This model can simulate many phenomena in the experiment qualitatively. Iwan and Blevins [10] and Iwan [77] deduced the coupling equation of wake oscillator and structural motion based on momentum conservation principle. The model has clear physical meaning and well reflects the hydrodynamic characteristics of VIV. The empirical parameters in the equation are determined by the test results of forced vibration. Therefore, the wake oscillator model is highly dependent on the empirical coefficient, and the empirical parameters selected by different wake oscillator models are greatly different. Therefore, even for the same research object, the prediction results of different wake oscillators are also significantly different. As demonstrated by Chaplin et al. [11], the amplitude response predicted by fourteen empirical models for the prediction of VIV in the same physical experiment is smaller than the experimental value, and the prediction results of each empirical model are also significantly different. The wake oscillator model has been widely accepted as a mainstream semi-empirical and semi-theoretical algorithm for a long time. However, the inherent deficiency of the algorithm lies in that the hydrodynamic frequency of the algorithm is obtained mainly through the calculation of the St number (St) obtained by the experiments of fixed circular cylinder. However, the frequency of hydrodynamic force of circular VIV of cylinder differs greatly from the frequency corresponding to the St number in the flow around the fixed cylinder. Therefore, in most cases, this algorithm can only obtain qualitative results, and its accuracy is greatly different from the experimental results. Facchinetti et al. [12] improved the dynamic characteristics of the wake oscillator model and, respectively, considered the coupling effect of displacement, velocity and acceleration on the wake oscillator. By comparing the prediction results of different coupling modes with the experimental results, it is found that the coupling of acceleration and lift induced by vortices can reflect the VIV characteristics of rigid cylinders quantitatively. Facchinetti et al. [13] further extended the model and applied it to VIV response of slender flexible cables. Mathelin and Langre [14] extend the work of Facchinetti et al. [13] to predict VIV response in shear flow. Furnes and Sørensen [15], Ge et al. [16] and Li et al. [17] proposed the simulation of dual-coupling oscillator model to simulate flow direction and transverse pulse dynamics, but the wake oscillator processed separately in both directions could not accurately predict the phase difference between flow direction and transverse response. Srinil [18] and Srinil and Zanganeh [19] used a double duffing-van der Pol oscillator to predict VIV responses to flow and lateral coupling. The model can predict the amplitude response of flow direction and transverse VIV more successfully. In general, the empirical model approach does not consider the specific flow field structure, but directly considers the flow field and structure as a whole system. A set of equations is used to describe the characteristics of the whole system, which has the advantages of simple model, wide application range, and low requirement for computing power and storage capacity of the computer. However, this set of equations has many parameters, and the selection of these parameters is determined by experience or experiment. The choice of these parameters is crucial to the result, and the choice of parameters of different empirical models varies greatly. How to apply forced and self-excited vibration test data under specific conditions has not been completely solved, which constitutes a major shortcoming of the empirical model method. In addition, the empirical model is mainly applied to the prediction of VIV response of riser, which is not suitable for the study of VIV mechanism, wake vortex law and fluid-solid coupling characteristics of riser.

3. Numerical model based on CFD method

3.1 Numerical simulation of 2D VIV

Numerical simulation of 2D VIV can be divided into two aspects. On the one hand, how to simulate the numerical results and phenomena which match the physical experiments; on the other hand, the intrinsic mechanism of VIV is studied by numerical simulation. In the study of a cylinder VIV, there are two topics that attract most attention: one is the super upper branch, and another is phase jump.

3.1.1 Super upper branch

Jauvtis and Williamson [20, 21] studied the VIV characteristics of a cylindrical structure at low mass and damping with 2 DOF and found that when the mass ratio m^* dropped to 2.6, the 2 DOF system got a super upper branch of transverse response amplitude, the largest response amplitude reaching 1.5 D, which all previous experiments had failed to get, and the "2T" vortex mode corresponding to the maximum transverse amplitude is observed.

Blevins and Coughran [22] adopted the physical experiment method to study comprehensively the VIV of cylindrical structure for 1 DOF and 2 DOF with variable mass ratio, damping, and found that the measured maximum amplitude was 1.75 D for a smooth 2 DOF cylinder at Reynolds number of 139,000. Based on experimental study, many scholars have carried out numerical study of VIV. Most scholars have used 2D numerical simulation to study the VIV of a cylindrical structure with low mass-damping [23–26]. For Re = 100–200, numerical simulations [23–26], including our previous research [27–29], gave similar amplitude results ($A_y/D = 0.6$) which were lower than the expected values. Subsequently, some scholars simulated the VIV of a cylindrical structure with Re = 1000, with the amplitude of transverse vibration reaching a value of $A_y/D = 0.7$. All these numerical simulations were carried out at low Reynolds numbers, well below the Reynolds numbers used in the classic experiment [21]. For this reason, numerical simulation seemed to capture only the initial and lower branches, whereas the upper and super upper branches were absent. Recently, some scholars began to use numerical simulation methods to study the VIV of a cylindrical structure at Reynolds numbers matched with the classic experimental values. Guilmineau and

Queutey [30] used the incompressible two-dimensional Reynolds-Averaged Navier-Stokes (RANS) scheme to simulate the VIV of an elastically mounted rigid cylinder with low mass-damping, constrained to oscillate transversely to a free stream and compared their results with compared with the 1996 experimental results of Khalak and Williamson [31]. According the initial condition used, the simulations predict correctly the maximum amplitude. On the other hand, the numerical results do not match the upper branch found experimentally. However, these results are encouraging, because no simulations have yet predicted such a high amplitude of vibration. Wanderley et al. [32] used the Roe-Sweby scheme to solve the compressible RANS equations to simulate the VIV of an elastically mounted rigid cylinder for 1 DOF with the mass ratio $m^* = 1.8$. The numerical results obtained in the present work agree remarkably well with experimental data obtained from the literature (Khalak and Williamson [31]) which the mass ratio was equal to 2.4 and captured the corresponding response branch and vortex patterns. Pan et al. [33] also adopted the incompressible two-dimensional RANS scheme to simulate the VIV of an elastically mounted rigid cylinder for 1 DOF with the mass ratio $m^* = 2.4$, and the result was compared with the experimental data reported by Khalak and Williamson [31]. The absence of the upper branch in RANS simulations is explained in depth because of discrepancies, which exist between experiments and RANS simulations. Srinil et al. [34] presents an experimental and numerical investigation of an elastically mounted rigid cylinder for 2 DOF with variable nature frequency ratios fx/fy and reported that when fx/fy =1.0 and mass ratio $m^* = 3.5$, the transverse amplitude of the numerical simulation was much smaller than that of the experimental value. Gsell et al. [35] investigated the VIVs of an elastically mounted circular cylinder using a direct numerical simulation method. In the upper branch, the maximum amplitude predicted by the simulation at $U^* = 6.5$ (about 1.2D) is lower than the amplitude measured experimentally [21] at the same reduced velocity. The author thinks that some deviations are expected due to the difference in the value of Re. Zhao and Cheng [36, 37] used the incompressible two-dimensional Reynolds-Averaged Navier-Stokes scheme to obtain the super upper branch and 2T vortex shedding pattern with initial conditions of velocity increasing constantly, which was in a good agreement with the experimental results [21]. Li et al. [38] adopted two typical turbulent models to simulate the VIV of a cylindrical structure for

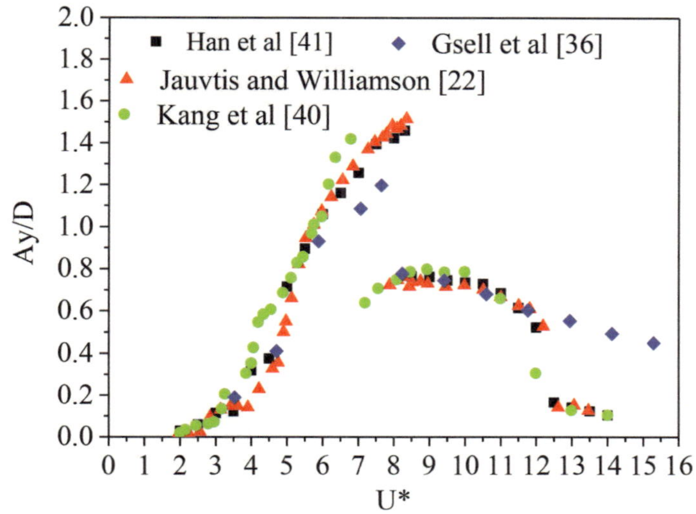

Figure 2.
Response amplitudes of the transverse directions under different reduced velocities [40].

1 DOF, and the predicted maximum amplitude was lower than that obtained by Khalak and Williamson [31]. Kang et al. [39] applied a modified SST model based on OpenFOAM to simulate experimental results [21] and it captured the maximum amplitude reaching values of 1.4 D when the entrance velocity was made to increase constantly in the process of numerical simulation. In previous studies [36, 37, 39], the initialization of numerical simulation started from the entrance, the inflow velocity is increased gradually from 0 or one low velocity value to the target value at a constant acceleration value and then is kept constant. The results show that the maximum transverse amplitude can be captured more accurately with appropriate inflow acceleration value. The value of this acceleration is usually very small, it takes a long time to accelerate to the target value, and how to determine the appropriate value of acceleration needs to be tested. Using this method to simulate the upper branch, the workload and computation time of is very large. Han et al. [40] successfully capture the initial branch, the lower branch, and the super upper branch with more accurate results, as shown in **Figure 2**. The units of each coordinate axis in the figure are dimensionless. The corresponding reduced velocity range of each branch is consistent with the classical experimental results [21]. The maximum value of the super upper branch is 1.46 D.

The vortex pattern at different reduced velocities is simulated successfully, as shown in **Figure 3**. The VIV of the streamwise direction is in a resonance state at a low reduced velocity and the vortex pattern in the wake is a symmetric pair in vortex modes. With the increase in reduced velocity, the transverse amplitude increases continuously, the cylindrical structure response enters the super upper branch, and the vortex pattern switches from 2S to 2T. When the reduced velocity increases further, the cylindrical structure's response enters the lower branch and the vortex pattern becomes 2P. With further increase in the reduced velocity, the transverse amplitude decreases continuously, and the vortex pattern becomes 2S.

3.1.2 Phase jump

Previous work has also shown phase differences between lift and CF displacements. In 1964, Bishop and Hassan [8] experimented on the lateral forced oscillation of a cylinder in a uniform flow, identifying the important phenomenon that when a cylinder oscillation frequency is near a natural vortex shedding frequency, the phase difference of the cylinder lift and the CF response undergoes a "sudden" jump from an "out-of-phase" mode to an "in-phase" mode. Then, Sarpkayab [41, 42], Bearman and Currie [43], Gopalkrishnan [44], and Carberry et al. [45] conducted similar experiments and reported the same phenomenon. The experimental study of Carberry et al. showed that in the process of a cylinder oscillating from a low frequency area (lower than the natural vortex shedding frequency) to a high frequency area (higher than the natural vortex shedding frequency), there is a transition zone. In the transition zone the mode of the vortex wake changes and at the same time the phase difference of the lift and CF responses undergoes a "sudden" jump. Sarpkaya [41] concluded that the phase difference between the lift and CF responses was closely related to an energy transfer between the cylinder and fluid. Zdravkovich [46] used visual means to analyze the previously studied flow field and found that a phase difference between the lift and CF responses was related to the vortex shedding time. Ongoren and Rockwell [47] used visualization analysis and reached a similar conclusion. Gu et al. [48] pointed out that with an increase in the frequency ratio, at a certain critical frequency, compared with the initial time, the cylindrical vortex shedding jumped from one side to another. Previous research has also shown that a phase difference between the cylinder lift and displacement, which determines

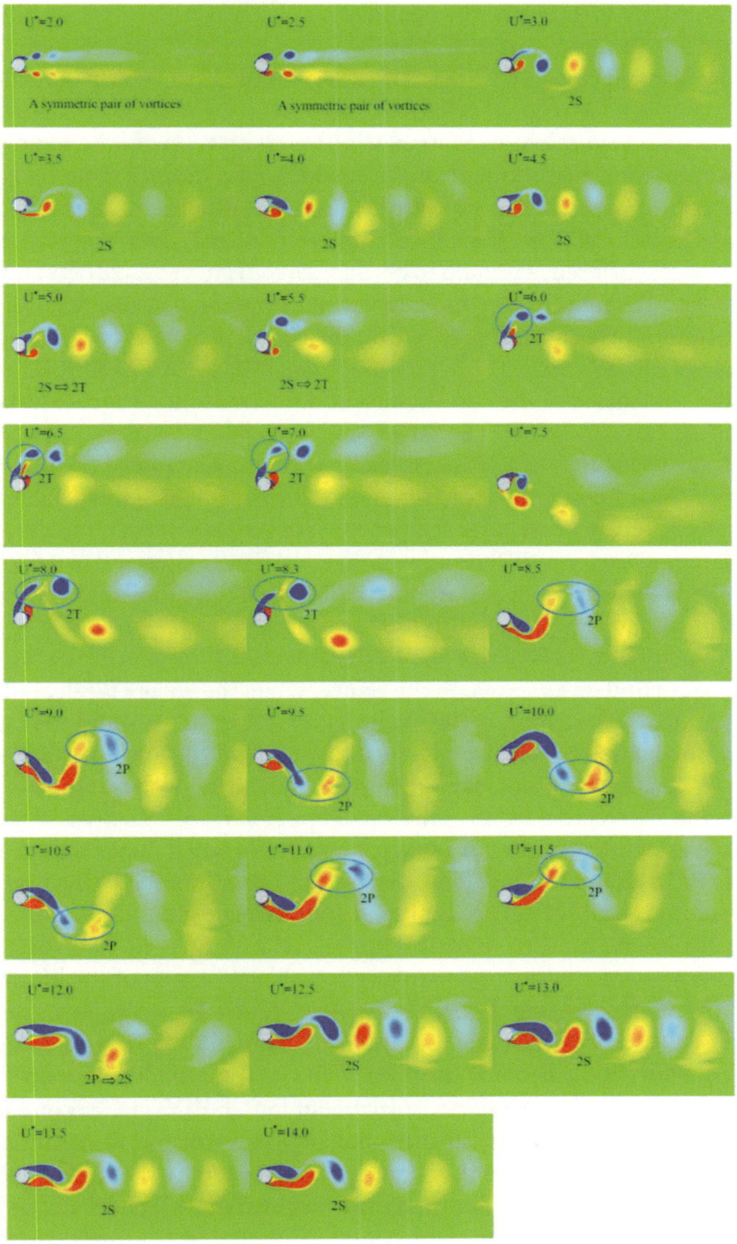

Figure 3.
Vorticity contours under different reduced velocity [40].

the amplitude of a cylinder vibration, is related to the energy transfer between the cylinder and fluid. Govardhan and Williamson [49] used the particle image velocimetry method to study transverse VIVs and highlighted that the phase remained slightly above throughout the initial and upper branch regimes and jumps of almost occurred throughout the upper and lower branch. Guilmineau and Queutey [30] used the RANS method to simulate a cylindrical CF forced oscillation and also found that the phase difference jumped about 180°. Li et al. [50] studied the nonlinear characteristics of VIV at and successfully captured phase-switch, lock-in, and beat phenomena. Wang et al. [51] has demonstrated the importance higher harmonic flow forces and phase mechanisms related to relative velocity and has study the effects of the relative velocity of the cylinder's oscillation with respect to the follow. Vortex formation modes been studied by 2D and 3D simulations before [6, 52–56]. The abovementioned research [6, 46–50, 52–56] highlighted typical behaviors in the relation between CF responses and lift and vortex shedding modes of the both elastically mounted cylinders and forced VIVs but did not clarify the science behind the law of the relationship between vortex shedding and phase difference, or how the vortex affects the phase difference. Despite numerous studies of typical behaviors of the VIV of an elastic cylinder, the underlying mechanisms governing typical behaviors such as phase difference remain to be elucidated. Han et al. [28] investigates the VIV of an elastically mounted cylinder at various frequency ratios. According to differences in the vortex shedding location, the vortex wake can be characterized by two kinds of mode, that is, the "first mode" and the "second mode". The mechanisms behind the phases of the first mode and the second mode vortex wakes are investigated and it is found that the flow speed induced by a cylindrical transverse vibration and the position of a vortex release are the root causes of the phase difference between the lift coefficient and transverse displacement. The speeds caused by a cylinder vibration and a cylinder shed vortex are the reasons that the lift amplitude of an oscillatory cylinder is different from that of a fixed cylinder. For a CF VIV, when a cylinder sheds the vortex, in addition to producing a flow similar to that of a fixed cylinder with a velocity of Δv_0, it also produces flow caused by the cylinder vibrating with a velocity of Δv_1. **Figure 4** shows the schematic diagram of the flow field around a fixed cylinder. As shown in the figure, when the vortex is released from the upper surface of the cylinder, flow velocity v is generated around the cylinder, making the upper surface velocity change to $U - \Delta v_0$ and the lower surface velocity change to $U + \Delta v_0$. At this point, the pressure on the upper surface is greater than the pressure on the lower surface due to the small upper surface velocity and the large small surface velocity, resulting in a downward lift. **Figure 5** shows the schematic diagram of the "first mode" flow field around a cylinder moving to the maximum displacement for the CF VIV. The "first mode" means when a cylinder moves to the maximum displacement, the cylindrical upper surface begins to shed vortex. At this time, Δv_1 and Δv_0 were in the opposite direction. The direction of the lift depended on the positive or negative condition of $\Delta v_0 - \Delta v_1$. When $\Delta v_0 - \Delta v_1 > 0$, the cylinder formed a downward lift and when $\Delta v_0 - \Delta v_1 < 0$, the cylinder formed an upward lift. **Figure 6** shows the schematic diagram of the "second mode" flow field around a cylinder moving to the maximum displacement for CF VIV. The "second mode" means when a cylinder moves to the maximal displacement and the cylindrical lower surface begins to shed vortex. At this time, Δv_1 and Δv_0 are in the same direction and the upper surface velocity is higher than the lower surface velocity, causing lowering of the cylinder's upper surface pressure compared to that of the lower surface pressure, thereby producing an upward lift [28].

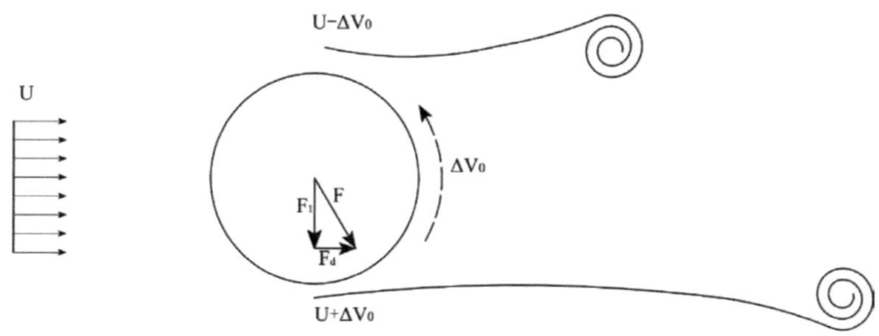

Figure 4.
Schematic diagram of the flow field around a fixed cylinder [28].

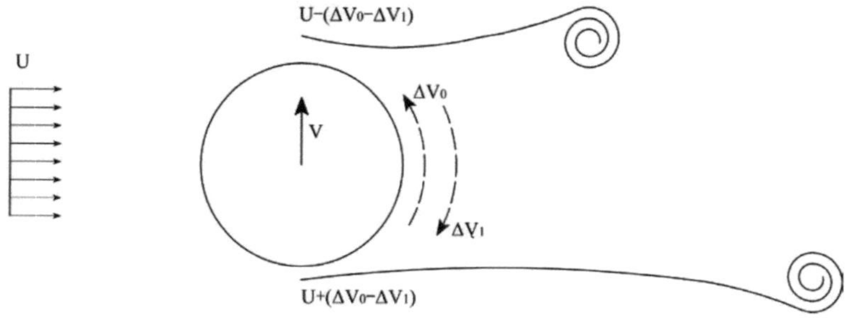

Figure 5.
Schematic diagram of the "first mode" flow field around a cylinder for CF VIV [28].

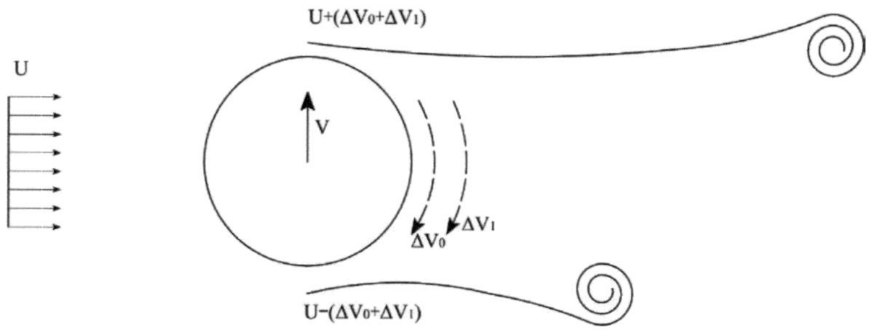

Figure 6.
Schematic diagram of the "second mode" flow field around a cylinder for CF VIV [28].

3.2 Numerical simulation of 2.5D VIV

The basic idea of the multi-strip method is to take a certain number of 2D slices of the riser in the axial direction to simulate the flow field and extract the load acting on the riser. Then, the load is applied to the riser to obtain the motion response of the riser, and the load is applied to the riser to obtain the movement response of the riser, which is carried out repeatedly, and finally to predict the VIV of the riser. This multi-strip method can effectively reduce the calculation time, and many scholars [57–61] compare the prediction data obtained by the multi-strip method with the experimental data, and the results are relatively consistent. Willden and Graham [59] used the multi-strip method to construct the quasi-three-dimensional riser model to simulate the VIV of riser in the transverse direction under shear flow and found that the fluid controlled the vibration frequency of the structure by influencing additional mass. Yamamoto et al. [57] established a quasi-three-dimensional CFD model, solved the hydrodynamic force of each section with the discrete vortex method, and calculated the vibration response of marine riser based on the Euler Bernoulli beam theory. The vibration response of the riser at different reduction velocities was calculated and compared with the test results. It was found that there was a vortex in the form of "2P" in the larger amplitude and a vortex in the form of "2S" in the smaller amplitude. Based on the theory of slicing method and the radial basis function dynamic mesh technology. Professor Wan [62–66] led the team to use OpenFOAM to solve the fluid field at each section and used the structural finite element method to simulate the motion response of the riser. The effects of various parameters (such as mass ratio, tip pretension, flow velocity and flow profile) on VIV of a riser were investigated. However, the VIV of a marine riser is a 3D problem in nature, and it is impossible to consider the influence of the flow along the axial direction by using the multi-strip method, and the description of the vibration mechanism of multi-modal coexistence is relatively vague. In addition, for the slicing method, it needs some interpolation algorithms to deal with the relationship between the slices, but this interpolation algorithm does not have a uniform standard [67, 68]. Hovor et al. [69] studied the three-dimensional effect of the wake of oscillating cylinder and the correlation of the flow force along the axial direction and proposed that the three-dimensional flow field characteristics of the vertical tube VIV are closely related to its structure amplitude, and the three-dimensional effect of the wake limited the further increase of the structure response amplitude.

3.3 Numerical simulation of 3D VIV

In recent years, due to the improvement of computer hardware, several fully 3D numerical simulations of a marine riser VIV are generated. Constantinides et al. [67] used a finite element Navier-Stokes (NS) solver to study a high L/D riser model. This method overcame the shortcomings of the Q3D method and correctly estimated the 3D effect. The response amplitude of the numerical simulation was compared with experiment to verify the rationality of the algorithm and vortex shedding modes were briefly analyzed. Holmes et al. [68] used fully 3D CFD methods to simulate a straked riser VIV. The resulting solutions were compared with available experiment data on a 38 m long riser model based on RMS displacements. Xie et al. [70] used a finite-volume method to study the multi-mode VIV of a flexible circular cylinder. Huang et al. [71, 72] used a RANS method and an overlapping mesh technique to simulate the VIV of a riser with the length to diameter ratio of 482 in uniform flow and shear flow environments and compared numerical calculation results with the experiment to verify the accuracy of the calculation procedure. Bourguet et al. [73–76] used direct numerical simulation of the 3D incompressible NS equations and a beam model to simulate the VIV of a cylindrical tensioned beam with the aspect ratio of 200 in linear and exponential shear flows at a Reynolds number equal to 330. Phasing mechanisms between the IL and CF, mono- and multi-frequency, and lock-in of VIV were explained.

In recent studies, many researches have investigated the VIV of marine risers but most of their research has been limited to 2D or rigid cylinder VIV. For the 3D flexible riser VIV, researchers have mostly adopted the 2.5D method. Only a few studies have adopted a bidirectional fluid-structure interaction method to simulate the VIV of the riser. In such studies, the displacement of the riser was solved using a mode superposition method or the Euler-Bernoulli bending beam equations. Most of these studies have concentrated on quantification of the vibration response. Nevertheless, the frequency characteristics of the vibration response and the universal rule of vortex shedding modes and trajectories in the spanwise direction have been less addressed, so further research is needed. Meanwhile, some studies have focused only on the VIV of a riser in linear and exponential shear flows at low Reynolds numbers, and further study is needed for the VIV of a riser in uniform flows at higher Reynolds numbers.

4. Research focus and prospect

4.1 Study on the nonlinear mechanism of VIV

Studies on VIV in recent decades are often accompanied by the discovery of new phenomena and locked-in, jump, lower branch, upper branch, super upper branch, lower branch, and various wake vortex forms (such as 2P, 2T, P + S) are found in cylindrical VIV; in the study of 3D marine riser VIV, the phenomenon of traveling-standing wave interaction propagation, multi-modal vibration and high-order modal vibration was found. Scholars have been continuously exploring the conditions and internal mechanism of these new phenomena, but it is difficult to give a conclusion at present. The goal is to further understand the phenomenon of VIV and extract the internal logic relations of major factors affecting VIV, to establish a more reasonable prediction model to serve ocean engineering practice.

4.2 Improve the prediction model of marine riser VIV

The prediction model of marine riser VIV still needs to be improved. Different from the numerical simulation based on CFD and FEM, these prediction models applied to practical engineering should be of high computational efficiency and can be easily applied to the structural design of marine riser. Perfect forecast model mainly from the following two aspects, that is, one is using simple parameters reflect the basic characteristics of VIV problem and special phenomenon of VIV (jumping, the response of each branch, etc.), and the selection of its value cannot rely too much on user's personal experience, avoid parameters selection process cumbersome and lose operability. Another is building empirical models that can be validated on a broader scale. At present, although there are various models, and each of them can be verified well under certain conditions, there will be great differences in the calculation of back-to-back verification. No model has absolute advantage, and it is difficult for one model to match the results of multiple experiments.

4.3 Fluid-structure interaction model based on CFD and CSD

With the continuous improvement of computer operation speed, numerical methods have been developed rapidly, and now have become one of the main research methods in parallel with theoretical analysis and tank test. The fluid-solid coupling numerical methods based on CFD and CSD are playing more and more important roles in the field of ocean engineering, especially in the design of offshore platforms and structural safety reliability analysis. The development of efficient large-scale parallel numerical analysis system will be an important direction to enhance the competitiveness in the field of ocean engineering in the future.

The 3D numerical simulation of the VIV of marine riser requires a hydrodynamic computational program and a structural mechanical computational program. In the coupling interface, fluid dynamic data and structural displacement data are exchanged, which involves grid data mapping, interpolation and fluid mesh reconstruction of fluid and solid media, resulting in the reduction of precision. In addition, the computing technology also involves some key technical problems, such as efficient interactive use between different solvers, economic and practical parallel algorithm and parallel programming, etc., and exploring more effective algorithms to further improve precision and efficiency is still one of the focuses of recent research. With the increase of the length-diameter ratio of marine riser and the increase of Reynolds number (the actual Reynolds number of marine engineering is mostly within the range of 10^5–10^6), the above problems become more difficult to solve. Research on circular VIV within the range of high Reynolds number is still rare, so it is necessary to conduct research on VIV under the condition of high Reynolds number, further enrich the research results, explore new fluid-solid coupling phenomena and theories, and lay a theoretical foundation for the conceptual design of marine engineering riser.

5. Conclusions

This chapter summarizes the research status of various numerical calculation methods, including empirical model and CFD numerical calculation. In the aspect of experience model, this chapter briefs the development process of experience model and the deficiency of current experience model is proposed. In the aspect of 2D numerical simulation, the progress of 2D numerical simulation of VIV is summarized combining the author's own research work on the two hot phenomena of super upper branch and phase jump. The development process of the multi-strip method for 3D marine riser VIV simulation is introduced in the aspect of 2.5D numerical simulation. In the aspect of 3D numerical simulation, numerical simulation of 3D marine riser VIV is analyzed with the limitations of current 3D marine riser research. At the end of the chapter, the future research direction is proposed, including the nonlinear mechanism of VIV, prediction models, and CFD and CSD fluid-solid interaction models.

Acknowledgements

This study was supported by the Guangxi natural science foundation (No. 2018GXNSFBA281138), the National Natural Science Foundation of China (No. 51809144), the Guangxi Major science and technology projects (No. AA17292007), a middle-aged and young teachers' basic ability promotion project of Guangxi Zhuang Autonomous Region of China (2019KY0443), Qinzhou College Scientific Research Project (2016PY-SJ08), Key subject of Guangxi (Naval Architecture and Ocean Engineering of Qinzhou University) Foundation, and Haiou Talent Plan of Qinzhou City.

Conflict of interest

No conflict of interest declared.

Appendices and nomenclature

A_y	the amplitude of the transverse vibration
D	the cylinder diameter
U	the inlet velocity
$m^* = m/m_d$	ratio of oscillating mass over displaced mass
$m_a = \rho\pi D^2/4$	added mass in still water
$St = f_0 D/U$	Strouhal number in still water
$\mathrm{Re} = UD/\nu$	Reynolds number
ν	kinematic viscosity coefficient
$U^* = U/(f_n D)$	reduced velocity in water
$f_n = \frac{1}{2\pi}\sqrt{\frac{k}{m+m_a}}$	natural frequency of elastic cylinder in still water
f_{nx}	natural frequency of elastic cylinder of x direction in still water
f_{ny}	natural frequency of elastic cylinder of y direction in still water

Author details

Xiangxi Han[1,2,3*], Youhong Tang[4], Zhiqiang Feng[1,2], Zhanbin Meng[1,2], Ang Qiu[3,5], Wei Lin[3] and Jiaming Wu[3*]

1 Guangxi Ship Digital Design and Advanced Manufacturing Research Center of Engineering Technology, Qinzhou University, Guangxi, China

2 Qinzhou Key Laboratory of Marine Advanced Design and Manufacturing, Qinzhou University, Guangxi, China

3 Department of Naval Architecture and Ocean Engineering, School of Civil Engineering and Transportation, South China University of Technology, Guangdong, China

4 College of Science and Engineering, Flinders University, South Australia, Australia

5 Guangdong Sinoway Composites Co., Ltd., Guangdong, China

*Address all correspondence to: hanxiangxi@qzhu.edu.cn and ctjmwu@scut.edu.cn

References

[1] Williamson CHK. Vortex dynamics in the cylinder wake. Annual Review of Fluid Mechanics. 1996;**28**(1):477-539. DOI: 10.1146/annurev.fl.28.010196.002401

[2] Williamson CHK, Govardhan R. Vortex-induced vibrations. Annual Review of Fluid Mechanics. 2004;**36**(1):413-455. DOI: 10.1146/annurev.fluid.36.050802.122128

[3] Williamson CHK, Govardhan R. A brief review of recent results in vortex-induced vibrations. Journal of Wind Engineering and Industrial Aerodynamics. 2008;**96**(6-7):713-735. DOI: 10.1016/j.jweia.2007.06.019

[4] Sarpkaya T. A critical review of the intrinsic nature of vortex-induced vibrations. Journal of Fluids and Structures. 2004;**19**(4):389-447. DOI: 10.1016/j.jfluidstructs.2004.02.005

[5] Gabbai RD, Benaroya H. An overview of modeling and experiments of vortex-induced vibration of circular cylinders. Journal of Sound and Vibration. 2005;**282**(3-5):575-616. DOI: 10.1016/j.jsv.2004.04.017

[6] Bearman PW. Vortex shedding from oscillating bluff bodies. Annual Review of Fluid Mechanics. 2003;**16**(1):195-222. DOI: 10.1146/annurev.fl.16.010184.001211

[7] Wu X, Ge F, Hong Y. A review of recent studies on vortex-induced vibrations of long slender cylinders. Journal of Fluids and Structures. 2012;**28**(1):292-308. DOI: 10.1016/j.jfluidstructs.2011.11.010

[8] Bishop RED, Hassan AY. The lift and drag forces on a circular cylinder oscillating in a flowing fluid. Proceedings of the Royal Society of London. Series A. Mathematical and Physical Sciences. 1964;**277**(1368):51-75. DOI: 10.1098/rspa.1964.0005

[9] Hartlen RT, Currie IG. Lift-oscillator model of vortex-induced vibration. Journal of the Engineering Mechanics Division. 1970;**96**(5):577-591

[10] Iwan WD, Blevins RD. A model for vortex induced oscillation of structures. Journal of Applied Mechanics. 1974;**41**(3):581-586. DOI: 10.1115/1.3423352

[11] Chaplin JR, Bearman PW, Cheng Y, Fontaine E, JMR G, Herfjord K, et al. Blind predictions of laboratory measurements of vortex-induced vibrations of a tension riser. Journal of Fluids and Structures. 2005;**21**(1):25-40. DOI: 10.1016/j.jfluidstructs.2005.05.016

[12] Facchinetti ML, Langre ED, Biolley F. Coupling of structure and wake oscillators in vortex-induced vibrations. Journal of Fluids and Structures. 2004;**19**(2):123-140. DOI: 10.1016/j.jfluidstructs.2003.12.004

[13] Facchinetti ML, Langre ED, Biolley F. Vortex-induced travelling waves along a cable. European Journal of Mechanics-B/Fluids. 2004;**23**(1):199-208. DOI: 10.1016/j.euromechflu.2003.04.004

[14] Mathelin L, Langre E. Vortex-induced vibrations and waves under shear flow with a wake oscillator model. European Journal of Mechanics—B/Fluids. 2005;**24**(4):478-490. DOI: 10.1016/j.euromechflu.2004.12.005

[15] Furnes GK, Sørensen K. Flow induced vibrations modeled by coupled non-linear oscillator. The Seventeenth International Offshore and Polar Engineering Conference; 1-6 July 2007; Lisbon, Portugal. 2007. p. 2781-2787

[16] Ge F, Long X, Wang L, Hong YS. Flow-induced vibrations of long circular

cylinders modeled by coupled nonlinear oscillators. Science in China Series G Physics Mechanics and Astronomy. 2009;**52**(7):1086-1093. DOI: 10.1007/s11433-009-0128-8

[17] Li XM, Guo HY, Meng FS. Nonlinear coupled in-line and cross-flow vortex-induced vibration analysis of top tensioned riser. China Ocean Engineering. 2010;**24**(4):749-758

[18] Srinil N. Analysis and prediction of vortex-induced vibrations of variable-tension vertical risers in linearly sheared currents. Applied Ocean research. 2011;**33**(1):41-53. DOI: 10.1016/j.apor.2010.11.004

[19] Srinil N, Zanganeh H. Modelling of coupled cross-flow/in-line vortex-induced vibrations using double Duffing and van der Pol oscillators. Ocean Engineering. 2012;**53**(3):83-97. DOI: 10.1016/j.oceaneng.2012.06.025

[20] Jauvtis N, Williamson CHK. Vortex-induced vibration of a cylinder with two degrees of freedom. Journal of Fluids and Structures. 2003;**17**(7):1035-1042. DOI: 10.1016/S0889-9746(03)00051-3

[21] Jauvtis N, Williamson CHK. The effect of two degrees of freedom on vortex-induced vibration at low mass and damping. Journal of Fluid Mechanics. 2004;**509**:23-62. DOI: 10.1017/S0022112004008778

[22] Blevins RD, Coughran CS. Experimental investigation of vortex-induced vibration in one and two dimensions with variable mass, damping, and Reynolds number. Journal of Fluids Engineering. 2009;**131**(10):101202. DOI: 10.1115/1.3222904

[23] Blackburn HM, Govardhan RN, Williamson CHK. A complementary numerical and physical investigation of vortex induced vibration. Journal of Fluids and Structures. 2001;**15**(3-4): 481-488. DOI: 10.1006/jfls.2000.0345

[24] Brika D, Laneville A. Vortex-induced vibrations of a long flexible circular cylinder. Journal of Fluid Mechanics. 1993;**250**:481-508. DOI: 10.1017/S0022112093001533

[25] Newman DJ, Karniadakis GE. A direct numerical simulation study of flow past a freely vibrating cable. Journal of Fluid Mechanics. 1997;**344**(01):95-136. DOI: 10.1017/S002211209700582X

[26] Zhou CY, So R, Lam K. Vortex-induced vibrations of an elastic circular cylinder. Journal of Fluids and Structures. 1999;**13**(2):165-189. DOI: 10.1006/jfls.1998.0195

[27] Han X, Lin W, Tang Y, Zhao C, Sammut K. Effects of natural frequency ratio on vortex-induced vibration of a cylindrical structure. Computers & Fluids. 2015;**110**:62-76. DOI: 10.1016/j.compfluid.2014.12.022

[28] Han X, Zhang X, Tang Y, Qiu A, Lin W, Zhao C. Dynamic mechanism of phase differences in one degree-of-freedom vortex-induced vibration of a cylindrical structure. Proceedings of the Institution of Mechanical Engineers, Part M: Journal of Engineering for the Maritime Environment. 2017. DOI: 10.1177/1475090217717356

[29] Han X, Lin W, Zhang X, Tang Y, Zhao C. Two degree of freedom flow-induced vibration of cylindrical structures in marine environments: Frequency ratio effects. Journal of Marine Science and Technology. 2016;**21**(3):1-14. DOI: 10.1007/s00773-016-0370-5

[30] Guilmineau E, Queutey P. Numerical simulation of vortex-induced vibration of a circular cylinder with low mass-damping in a turbulent flow. Journal of Fluids and Structures. 2004;**19**(4):449-466. DOI: 10.1016/j.jfluidstructs.2004.02.004

[31] Khalak A, Williamson CH. Dynamics of a hydroelastic cylinder with

very low mass and damping. Journal of Fluids and Structures. 1996;**10**(5):455-472. DOI: 10.1006/jfls.1996.0031

[32] Wanderley JBV, Souza GHB, Sphaier SH, Levi C. Vortex-induced vibration of an elastically mounted circular cylinder using an upwind TVD two-dimensional numerical scheme. Ocean Engineering. 2008;**35**(14-15):1533-1544. DOI: 10.1016/j.oceaneng.2008.06.007

[33] Pan ZY, Cui WC, Miao QM. Numerical simulation of vortex-induced vibration of a circular cylinder at low mass-damping using RANS code. Journal of Fluids and Structures. 2007;**23**(1):23-37. DOI: 10.1016/j.jfluidstructs.2006.07.007

[34] Srinil N, Zanganeh H, Day A. Two-degree-of-freedom VIV of circular cylinder with variable natural frequency ratio: Experimental and numerical investigations. Ocean Engineering. 2013;**73**(8):179-194. DOI: 10.1016/j.oceaneng.2013.07.024

[35] Gsell S, Bourguet R, Braza M. One- versus two-degree-of-freedom vortex-induced vibrations of a circular cylinder at Re=3900. Journal of Fluids and Structures. 2016;**67**:156-172. DOI: 10.1016/j.jfluidstructs.2016.09.004

[36] Zhao M, Cheng L. Numerical investigation of local scour below a vibrating pipeline under steady currents. Coastal Engineering. 2010;**57**(4):397-406. DOI: 10.1016/j.coastaleng.2009.11.008

[37] Zhao M, Cheng L. Numerical simulation of two-degree-of-freedom vortex-induced vibration of a circular cylinder close to a plane boundary. Journal of Fluids and Structures. 2011;**27**(7):1097-1110. DOI: 10.1016/j.jfluidstructs.2011.07.001

[38] Li W, Li J, Liu S. Numerical simulation of vortex-induced vibration of a circular cylinder at low mass and damping with different turbulent models. In: Ocean; 7-10 April 2014; TAIPEI. New York: IEEE; 2014. pp. 1-7

[39] Kang Z, Ni W, Sun L. A numerical investigation on capturing the maximum transverse amplitude in vortex induced vibration for low mass ratio. Marine Structures. 2017;**52**:94-107. DOI: 10.1016/j.marstruc.2016.11.006

[40] Han X, Lin W, Wang D, Qiu A, Feng Z, Tang Y, et al. Numerical simulation of super upper branch of a cylindrical structure with a low mass ratio. Ocean Engineering. 2018;(15):108-120. DOI: 10.1016/j.oceaneng.2018.09.014

[41] Sarpkaya T. Fluid forces on oscillating cylinders. Journal of the Waterway, Port, Coastal and Ocean Division. 1978;**104**(3):275-290

[42] Sarpkaya T. Hydrodynamic damping, flow-induced oscillations, and biharmonic response. Journal of Offshore Mechanics and Arctic Engineering. 1995;**117**(4):232-238. DOI: 10.1115/1.2827228

[43] Bearman PW, Currie IG. Pressure-fluctuation measurements on an oscillating circular cylinder. Journal of Fluid Mechanics. 1979;**91**(04):661-677. DOI: 10.1017/S0022112079000392

[44] Gopalkrishnan R. Vortex-induced forces on oscillating bluff cylinders [thesis]. Cambridge: Massachusetts Institute of Technology; 1993. p. 1993

[45] Carberry J, Sheridan J, Rockwell D. Controlled oscillations of a cylinder: Forces and wake modes. Journal of Fluids & Structures. 2005;**538**:31-69. DOI: 10.1017/S0022112005005197

[46] Zdravkovich MM. Modification of vortex shedding in the synchronization range. Journal of Fluids Engineering. 1982;**104**(4):513-517. DOI: 10.1115/1.3241895

[47] Ongoren A, Rockwell D. Flow structure from an oscillating cylinder. I—Mechanisms of phase shift and

recovery in the near wake. II—Mode competition in the near wake. Journal of Fluid Mechanics. 1988;**191**:197-223. DOI: 10.1017/S0022112088001569

[48] Gu W, Chyu C, Rockwell D. Timing of vortex formation from an oscillating cylinder. Physics of Fluids. 1994;**6**(11): 3677-3682. DOI: 10.1063/1.868424

[49] Govardhan R, Williamson CHK. Modes of vortex formation and frequency response of a freely vibrating cylinder. Journal of Fluid Mechanics. 2000;**420**:85-130. DOI: 10.1017/S0022112000001233

[50] Li T, Zhang J, Zhang W. Nonlinear characteristics of vortex-induced vibration at low Reynolds number. Communications in Nonlinear Science and Numerical Simulation. 2011;**16**:2753-2771. DOI: 10.1016/j.cnsns.2010.10.014

[51] Wang XQ, Rmc S, Chan KT. A non-linear force model for vortex-induced vibration of an elastic cylinder. Journal of Sound and Vibration. 2003;**260**(2):287-305. DOI: 10.1016/S0022-460X(02)00945-8

[52] Evangelinos C, Karniadakis GE. Dynamics and flow structures in the turbulent wake of rigid and flexible cylinders subject to vortex-induced vibrations. Journal of Fluid Mechanics. 1999;**400**:91-124. DOI: 10.1017/S0022112099006606

[53] Lucor D, Foo J, Karniadakis GE. Vortex mode selection of a rigid cylinder subject to VIV and low mass damping. Journal of Fluids and Structures. 2005;**20**:483-503. DOI: 10.1016/j.jfluidstructs.2005.02.002

[54] Williamson CHK. Vortex dynamics in the cylinder wake. Annual Review of Fluid Mechanics. 2003;**28**:477-539. DOI: 10.1146/annurev.fl.28.010196.002401

[55] Williamson CHK, Roshko A. Vortex formation in the wake of oscillating cylinder. Journal of Fluids and Structures. 1988;**2**:355-381. DOI: 10.1016/S0889-9746(88)90058-8

[56] Zhao M, Cheng L, An H, Lu L. Three-dimensional numerical simulation of VIV of an elastically mounted rigid circular cylinder in steady current. Journal of Fluids and Structures. 2014;**50**:292-311. DOI: 10.1016/j.jfluidstructs.2014.05.016

[57] Yamamoto CT, Fregonesi RA, Meneghini JR, Saltara F. Numerical simulation of the flow around flexible cylinders. In: 21st International Conference on Offshore Mechanics and Arctic Engineering; 23-28 June 2002; Oslo. 2002. pp. 837-846

[58] Shulz KW, Meling TS. Multi-strip numerical analysis for flexible riser response. In: 23rd International Conference on Offshore Mechanics and Arctic Engineering; 20-25 June 2004; Vancouver. 2004. pp. 379-384

[59] Willden RHJ, Graham JMR. Multi-modal vortex-induced vibrations of a vertical riser pipe subject to a uniform current profile. European Journal of Mechanics-B/Fluids. 2004;**23**(1):209-218. DOI: 10.1016/j.euromechflu.2003.09.011

[60] Willden RHJ, Graham JMR. CFD simulations of the vortex-induced vibrations of model riser pipes. In: 24th International Conference on Offshore Mechanics and Arctic Engineering; 12-17 June 2005; Halkidiki. 2005. pp. 837-846

[61] Willden RHJ, Graham JMR. Numerical prediction of VIV on long flexible circular cylinders. Journal of Fluids and Structures. 2001;**15**(15):659-669. DOI: 10.1006/jfls.2000.0359

[62] Fu B, Zou L, Wan D. Numerical study of vortex-induced vibrations of a flexible cylinder in an oscillatory flow. Journal of Fluids & Structures. 2018;**77**:170-181. DOI: 10.1016/j.jfluidstructs.2017.12.006

[63] Fu B, Wan D. Numerical study of vibrations of a vertical tension riser excited at the top end. Journal of Ocean Engineering and Science. 2017;**2**(4):268-278. DOI: 10.1016/j.joes.2017.09.001

[64] Yu D, Lu Z, Wan D. Numerical simulations of vortex-induced vibrations of a flexible riser with different aspect ratios in uniform and shear currents. Journal of Hydrodynamics, Ser. B. 2017;**29**(6):1010-1022. DOI: 10.1016/S1001-6058(16)60815-6

[65] Yu D, Zou L, Wan D. Numerical analysis of multi-modal vibrations of a vertical riser in step currents. Ocean Engineering. 2017;**152**:428-442. DOI: 10.1016/j.oceaneng.2017.12.033

[66] Ye H, Wan D. Benchmark computations for flows around a stationary cylinder with high Reynolds numbers by RANS-overset grid approach. Applied Ocean Research. 2017;**65**:315-326. DOI: 10.1016/j.apor.2016.10.010

[67] Constantinides Y, Oakley OH, Holmes S. CFD high L/D riser modeling study. In: 26th International Conference on Offshore Mechanics and Arctic Engineering; 10-15 June 2007; San Diego. 2007. pp. 715-722

[68] Holmes S, Oakley OH, Constantinides Y. Simulation of riser VIV using fully three dimensional CFD simulations. 25th International Conference on Offshore Mechanics and Arctic Engineering; 4-9 June 2006; Hamburg. 2006. pp. 563-570

[69] Hover FS, Davis JT, Triantafyllou MS. Three-dimensionality of mode transition in vortex-induced vibrations of a circular cylinder. European Journal of Mechanics-B/Fluids. 2004;**23**:29-40. DOI: 10.1016/j.euromechflu.2003.04.002

[70] Xie F, Deng J, Zheng Y. Multi-mode of vortex-induced vibration of a flexible circular cylinder. Journal of Hydrodynamics, Series B. 2011;**23**:483-490. DOI: 10.1016/S1001-6058(10)60139-4

[71] Huang K, Chen HC, Chen CR. Vertical riser VIV simulation in uniform current. 28th International Conference on Offshore Mechanics and Arctic Engineering; 31 May–5 June, 2009 Honolulu. 2009. pp. 395-405

[72] Huang K, Chen HC, Chen CR. Vertical riser VIV simulation in sheared current. 19th International Offshore and Polar Engineering Conference; 21-26 June 2009 Osaka. 2009. pp. 1369-1376

[73] Bourguet R, Karniadakis GE, Triantafyllou MS. Lock-in of the vortex-induced vibrations of a long tensioned beam in shear flow. Journal of Fluids & Structures. 2011;**27**:838-847. DOI: 10.1016/j.jfluidstructs.2011.03.008

[74] Bourguet R, Lucor D, Triantafyllou MS. Mono- and multi-frequency vortex-induced vibrations of a long tensioned beam in shear flow. Journal of Fluids & Structures. 2012;**32**:52-64. DOI: 10.1016/j.jfluidstructs.2011.05.008

[75] Bourguet R, Karniadakis GE, Triantafyllou MS. Multi-frequency vortex-induced vibrations of a long tensioned beam in linear and exponential shear flows. Journal of Fluids & Structures. 2013;**41**:33-42. DOI: 10.1016/j.jfluidstructs.2012.07.007

[76] Bourguet R, Karniadakis GE, Triantafyllou MS. Phasing mechanisms between the in-line and cross-flow vortex-induced vibrations of a long tensioned beam in shear flow. Computers & Structures. 2013;**122**:155-163. DOI: 10.1016/j.compstruc.2013.01.002

[77] Iwan WD. The Vortex-induced oscillation of non-uniform structure analysis. Journal of Sound and Vibration. 1981;**79**:291-301. DOI: 10.1016/0022-460X(81)90373-4

Chapter 8

Propagation Buckling of Subsea Pipelines and Pipe-in-Pipe Systems

Hassan Karampour and Mahmoud Alrsai

Abstract

This chapter investigates buckle propagation of subsea single-walled pipeline and pipe-in-pipe (PIP) systems under hydrostatic pressure, using 2D analytical solutions, hyperbaric chamber tests and 3D FE analyses. Experimental results are presented using hyperbaric chamber tests, and are compared with a modified analytical solution and with numerical results using finite element analysis for single-walled pipelines and PIPs. The experimental investigation is conducted using commercial aluminum tubes with diameter-to-thickness (D/t) ratio in the range 20–48. The comparison indicates that the modified analytical expression presented in this work provides a more accurate lower bound estimate of the propagation buckling pressure of PIPs compared to the existing equations, especially for higher D_o/t_o ratios. A 3D FE model is developed and is validated against the experimental results of the propagation bucking. A parametric FE study is carried out and empirical expressions are provided for buckle propagation pressures of PIPs with (D_o/t_o) ratio in the range 15–25. Moreover, empirical expressions are proposed for the collapse pressure of the inner pipe (P_{ci}), the proposed empirical equation for P_{ci}, is shown to agree well with the experimental results of the tested PIPs.

Keywords: collapse pressure, external pressure, offshore pipelines, pipe-in-pipe, propagation buckling

1. Propagation buckling of single pipe

1.1 Introduction

Deep and ultra-deep water pipelines are vulnerable to propagation buckling due to the high external pressures. The pipeline may collapse due to the local dents, imperfections and ovalizations in the pipe-wall. This collapse will change the cross-section of the pipeline from a circular shape into a dog-bone or even flat shape. The buckle may then propagate along the pipeline and cause the pipeline to be shut down. A typical propagation buckle scenario is shown in **Figure 1**, which is triggered by impact on the pipeline from an anchor dropped from a passing vessel.

Different stages of the buckle are shown in **Figure 1** in terms of the external pressure versus change in volume of the pipe. The dent caused by the impact can initiate the buckle due to high external pressure. The elastic buckling is followed by a plastic collapse and change in the cross-section of the tube from circular to oval

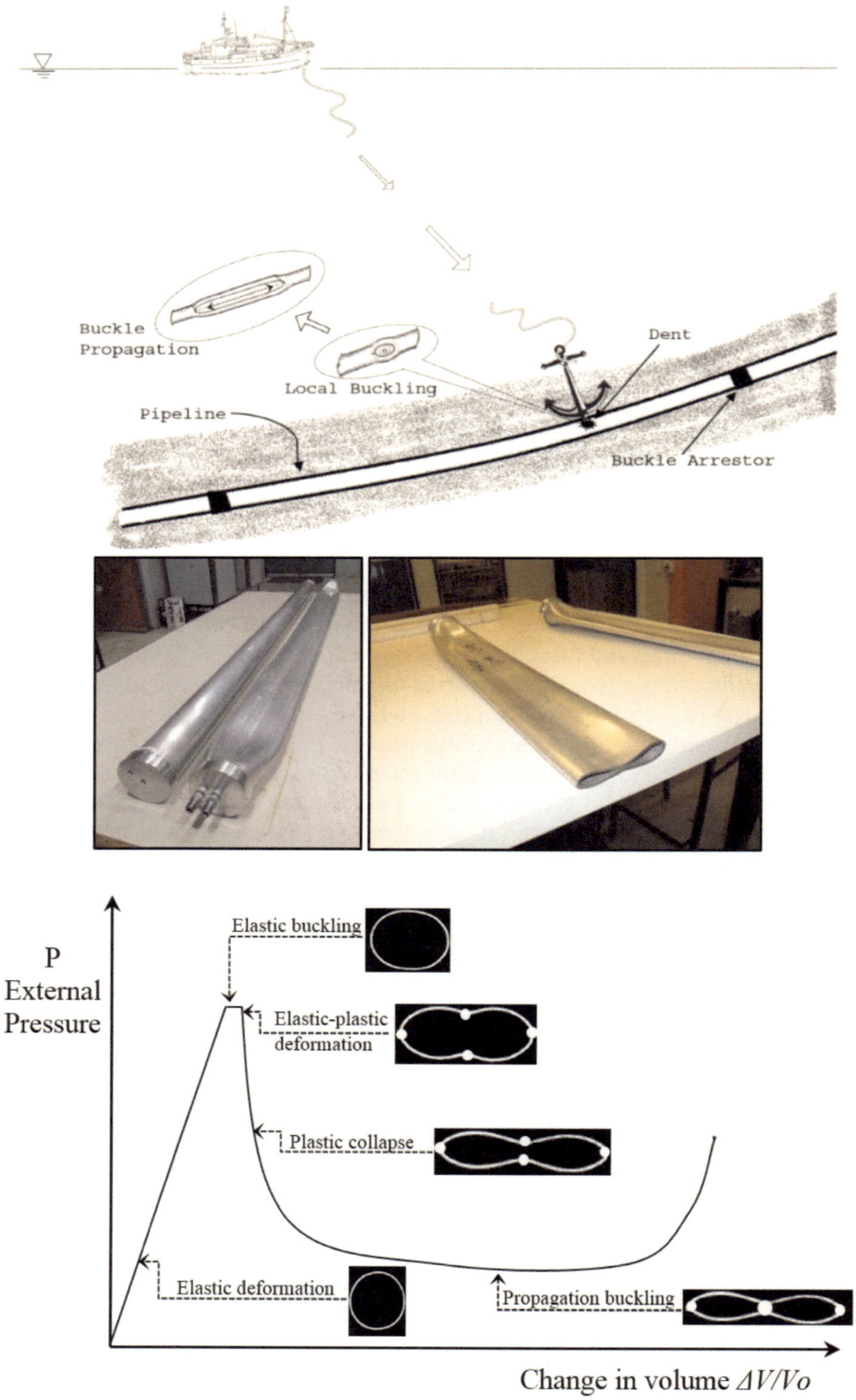

Figure 1.
Buckle propagation scenario [1].

and finally a dog-bone shape. If the pressure is maintained, the buckle will propagate quickly along the length of the pipe. Offshore pipelines normally experience high service external pressure; therefore the buckle will propagate through the length, forcing the flow line to be shut. The lowest pressure that maintains propagation is known as the propagation pressure, and is much smaller than the collapse pressure. To account for the difference between the collapse and the propagation pressures in design, a thick-walled pipeline is required [1, 2].

As shown in **Figure 1**, the propagation pressure is much less than initiation pressure (peak pressure in **Figure 1**). The initiation pressure is significantly affected by the size of the local dent. Local dents may also occur during the installation period. The most common types of offshore pipeline installation are S-lay method, J-lay method, Reel-lay method and Towing method. A combination of bending and external pressure happens in the sag bend length of the pipe. Normally high tension is applied to the pipe to maintain its stiffness during installation. If for any reason this tension is released, high bending in the sag bend region may cause local buckling which may be followed by propagation buckling. Apart from the foretold loading sources, manufacturing imperfections in pipe such as non-uniform thickness, varying elastic modulus, local ovalization, and also erosion and corrosion may cause local buckling in pipelines.

Many researchers have investigated various aspects of this problem since it was first presented by Mesloh et al. [3] and Palmer and Martin [4]. Most notably is the extensive work of Kyriakides [5, 6], Kamalarasa [7] and Albermani et al. [2]. Recent books by Kyriakides [1] and Palmer and King [8] provide comprehensive review of this problem and the associated literature. The work done by Xue et al. [9] investigates the effect of corrosion in the propagation buckling of subsea pipelines. Buckle arrestors [1, 18], pipe-in-pipe system [10–14], sandwich pipe system [15] and ring-stiffened pipelines [16], are used to confine the propagation buckling in subsea pipelines.

As stated before, a local dent or ovalization in the pipe wall can cause a local collapse as in the pipe-wall. It is well-known that the collapse pressure of a 2D arch (similar to a single pipeline (P_{cr})), under lateral pressure can be approximated by [17]:

$$P_{cr} = \frac{E}{4(1-\nu^2)} \left(\frac{t}{r}\right)^3 \tag{1}$$

where E is the modulus of elasticity, ν is the Poisson's ratio, t is the pipe wall thickness and r is the mean radius of pipe. As shown in **Figure 1** prior to the collapse pressure no significant change in cross section of pipe is observed. Note that for sake of clarity the slope of line ending to collapse pressure in **Figure 1** is exaggerated. During the propagation buckling the pipe endures substantial change in its shape.

1.2 Analytical solution of propagation pressure of single pipe

A typical buckle propagation response is characterized by the pressure at which the snap-through takes place (the initiation pressure P_I) and the pressure that maintains propagation (the propagation pressure P_p) which is a small fraction of P_I.

Palmer and Martin [4] suggested a 2D approximation for propagation buckling of subsea pipelines Eq. (2). Their solution is based on a 2D ring collapse (plane strain) mechanism, and accounts for the circumferential bending effect of the pipe wall (see **Figure 2**). The Palmer and Martin (PM) solution underestimates the propagation pressure when compared to experimental results. This difference increases as D/t decreases. The propagation pressure from the PM solution, P_{PM}, is given by:

$$P_{PM} = \frac{\pi}{4}\sigma_y\left(\frac{t}{r}\right)^2 \tag{2}$$

for a pipe with radius, r, wall thickness, t, and material yield stress, σy. Based on experimental observations from hyperbaric chamber tests, the top and bottom hinges in **Figure 2a** move towards each other while the left and right hinges move laterally away from each other. This deformation continues until touchdown (**Figure 2b**), the lateral movement seizes and flattening of the resulting four arch segments commence (**Figure 2c**).

Accordingly, a modification to the lower bound PM solution is proposed [2], by accounting for the circumferential membrane as well as flexural effects in the pipe wall

$$W_{ex} = (W_{in})_f + (W_{in})_m \tag{3}$$

where W_{ex} is the external work done by the net hydrostatic pressure and W_{in} is the internal work due to circumferential flexure, f, and membrane, m, effects. The initially circular cross section of the pipe (**Figure 2a**) will deform into a dog-bone (**Figure 2b**) and eventually into a nearly flat segment. Accordingly, (Eq. (3)) can be written as:

$$p(\Delta A) = 3\pi m_p + (pr)(\Delta l) \tag{4}$$

where ΔA is the change in the cross section area, Δl is the change in the circumferential length and $m_p$is the plastic moment, these are given by:

$$\Delta A = \pi r^2 \tag{5}$$

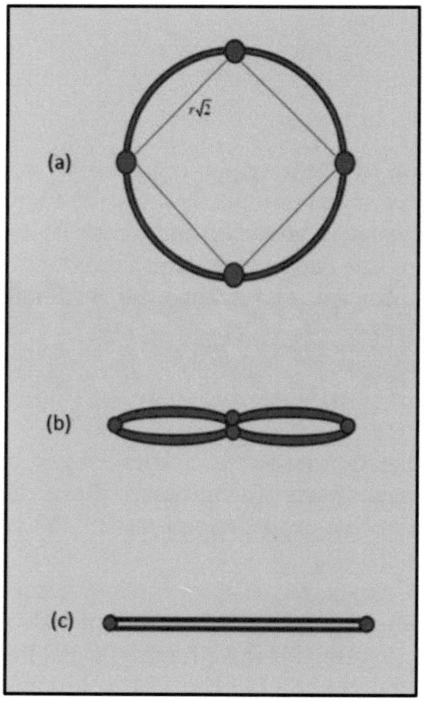

Figure 2.
A schematic of 2D deformation stages in propagation buckling of single pipe; (a) the initial circular cross section of the single pipe; (b) dog-bone deformed shape; (c) flat segment of the pipe.

$$\Delta l = 0.626r \tag{6}$$

$$m_p = \sigma_y \frac{t^2}{4} \tag{7}$$

Substituting Eqs. (5)–(7) into (4), the propagation pressure, $\widetilde{p}$, is obtained as:

$$\widetilde{p} = \frac{3}{2.515}\left[\frac{\pi}{4}\sigma_y\left(\frac{t}{r}\right)^2\right] = 1.193p_{PM} \tag{8}$$

Experimental observations confirm that the propagation pressure predicted by (Eq. (8)) is 19% higher than the PM prediction Eq. (2), regardless of t/r ratio. However, it should be noted, that by adopting plane strain conditions, the tensile coupon yield stress can be augmented by a factor of $(2/\sqrt{3})$ in (Eq. (8)) which results in an additional 15% increase in $\widetilde{p}$.

1.3 Experiments on propagation buckling of single-walled pipelines

A stiff 4 m long hyperbaric chamber rated for 20 MPa (2000 m water depth) internal pressure was used for testing (**Figure 3a**). Three meter long aluminum pipes were used in the hyperbaric chamber tests [2]. Ovalization measurements along the pipe samples before testing were carried out that gave an average ovalization ratio Ω (Eq. (9)) around 0.46–0.67%

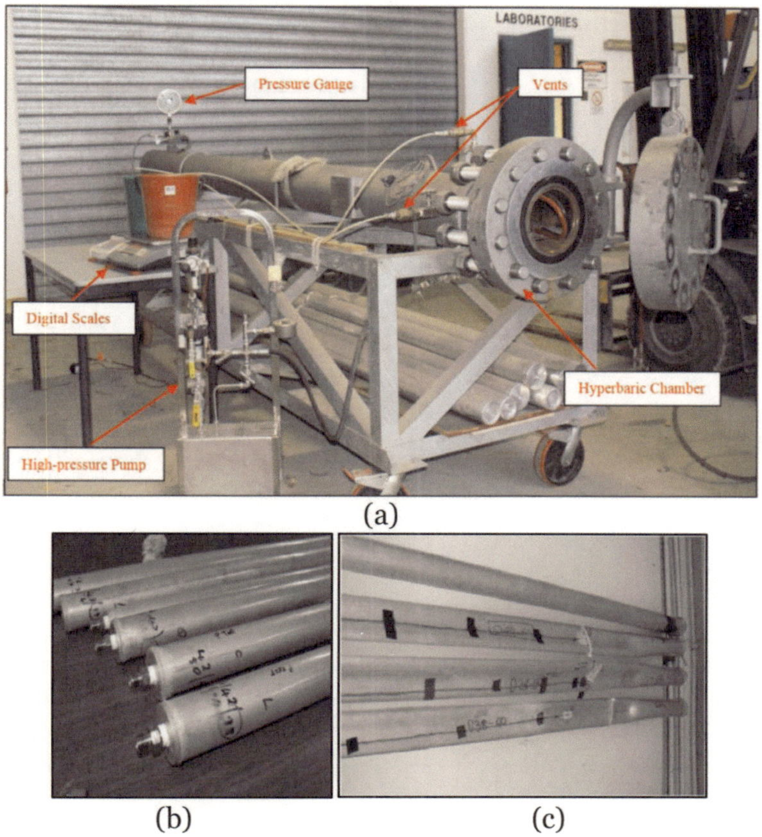

Figure 3.
The experimental set-up: (a) the hyperbaric chamber, high-pressure pump, scales, pressure gauge and vents, (b) pipes and fittings, (c) failed pipes tested in the hyperbaric chamber.

$$\Omega_0 = \frac{D_{\max} - D_{\min}}{D_{\max} + D_{\min}} \tag{9}$$

where D_{max} and D_{min} are the maximum and minimum measured outer diameters along the pipe length.

The hyperbaric chamber test procedure is as follows. Thick discs are welded at both ends of 3 m pipeline. The pipeline is then filled with water and inserted inside the chamber (**Figure 3b**). The bolts at the chamber lid are tightened using a pneumatic torque wrench and the chamber is sealed. Using a control-volume analogy, the water inside the chamber is pressurized at a slow rate, using a hand pump. When the pressure reaches the initiation pressure P_I of the pipeline, a section along the pipe sample collapses. This leads to a substantial drop in chamber pressure and is followed by water flowing from within the pipe sample through vent. Then, the chamber pressure is stabilized at the propagation pressure, P_p, with the buckle longitudinally propagating along the pipe sample accompanied by uniform water flow from the vent. The failed samples are sown in **Figure 3c**.

The average pressures of the 19 pipes tested in the hyperbaric chamber are represented in **Table 1**. A typical pressure-volume change response obtained from the hyperbaric chamber tests is shown in **Figure 4**. In **Figure 4**, the pressure inside the chamber is normalized by the propagation pressure, P_{PM}, and the change in the pipe volume ΔV is normalized by the initial volume of the pipe, V. As stated before, the buckle initiation pressure, P_I, is sensitive to imperfections (such as a dent in the pipe wall). However, the buckle propagation pressure, P_p, is not affected by the imperfection.

The analytical, experimental and numerical pressures are compared in **Table 2**. The ratio of propagation pressure from the hyperbaric chamber tests P_P to the

Sample/material		D/t	Coupon tests			Analytical (MPa)		Hyperbaric chamber (MPa)			
ID	Al-6060		σ_Y (MPa)	E/σ_Y	E'/E(%)	P_{PM} Eq. (2)	$\tilde{P}$ Eq. (8)	Experiment		Finite element	
								P_I	P_P	P_{IFE}	P_{PFE}
D50	T591	25	122	440	1.5	0.778	0.93	6.42	1.6	5.12	1.1
D60	T4	20	81	716	1.9	1.011	1.21	8.24	2.3	8.15	1.6
D76	T5	47.5	156	367	0.4	0.205	0.245	1.32	0.35	1.07	0.3

Table 1.
Summary of experimental, analytical and numerical results.

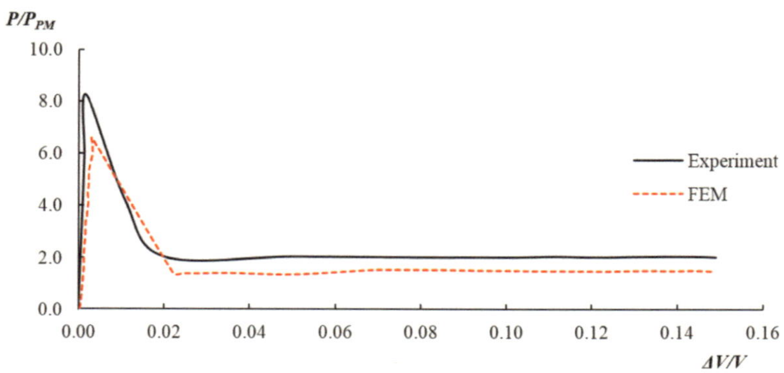

Figure 4.
Normalized pressure-volume response (experimental and numerical results) for D50.

Sample	D/t	Hyperbaric chamber		Finite element	
		$P_P/\widetilde{P}$	P_I/P_P	P_I/P_{IFE}	P_P/P_{PFE}
D50	25	1.720	4.01	1.253	1.453
D60	20	1.900	3.58	1.011	1.437
D76	47.5	1.428	3.77	1.234	1.167

Table 2.
Comparison of experimental, analytical and numerical results.

modified analytical solution $\tilde{p}$ (Eq. (8)) vary from 1.428 to 1.9 depending on *D/t* ratio. Mesloh et al. [19] suggested similar relations. The ratio of P_I/P_P from the hyperbaric chamber tests is also shown in **Table 2** and varies from 3.5 to 4.0. The results represented in **Table 2** highlight the susceptibility of deep and ultra-deep subsea pipelines to propagation buckling. To confine the buckle propagation, external ring stiffeners are exploited intermittently on the pipeline. These buckle arrestors can only confine the pressure between two stiffeners.

1.4 Finite element study on propagation buckling of single-walled pipelines

FE models were created in ANSYS [20] to investigate the response of the pipe to propagation buckling. Thin 4-noded shell elements (181) were used to model the pipe. SHELL181 is suitable for analyzing thin to moderately-thick shell structures. It is a four-noded element with six degrees of freedom at each node: translations in the x, y, and z directions, and rotations about the x, y, and z axes. Hydrostatic pressure can be applied as surface loads on corresponding surface. Pipe wall thickness is defined using section data command. A convergence study was performed and five integration points was found to be adequate for propagation buckling of cylindrical pipes. Frictionless contact and target elements (ANSYS elements 174 and 170) are used to define the contact between the inner surfaces of the pipe wall. These elements are created on the surface of the existing shell elements using ESURF command. The 3D contact surface elements CONTA174 are associated with the 3D target segment elements TARGE170 via a shared real constant set. Contact stiffness can be controlled by normal penalty stiffness factors and tangent penalty stiffness factor. Normal penalty stiffness factor of 0.1 was selected based on a convergence study performed that ensures both real contact behavior and reasonable computational time. Tangent stiffness factor appeared not to affect the results significantly.

A von-Mises elastoplastic material definition with isotropic hardening was adopted based on material properties shown in **Table 1**. Total of 40 shell-181 elements in circumference were utilized for modeling the pipe. Local ovalizations were introduced to FE model by applying external pressures symmetrically on 8 elements on top of the pipe along a length equal to diameter of the pipe. Geometry is then updated using UPGEOM command and nonlinear geometric and material analysis is carried out. The FE model is 3 m long and is restrained against translation at all nodes at both ends.

The initiation and propagation pressures obtained from FE analysis (P_{IFE} and P_{PFE} respectively) are summarized in **Table 1** and are in reasonable agreement with the experimental results from the hyperbaric chamber. Unlike buckle initiation pressure (P_I), buckle propagation pressure (P_P) is independent of curvature or ovalization of pipe. Palmer and Martin prediction P_{PM} estimates a lower bound for propagation pressure. The FE predictions of initiation and propagation pressures on

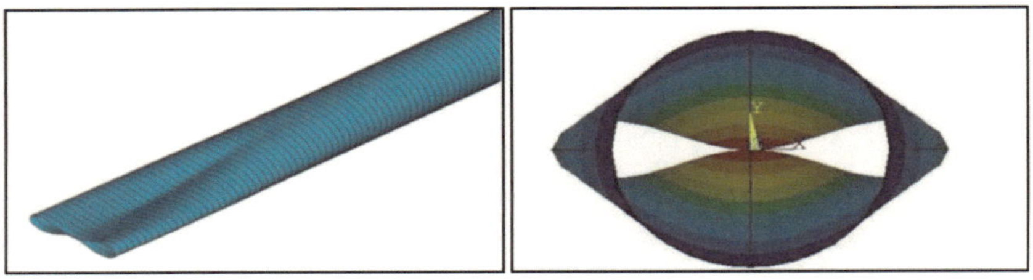

Figure 5.
FE model of 3 m long D50 showing the onset of propagation buckling.

average represent 87 and 74%, respectively, of the experimental results. A typical FE result for D50 pipe is shown in **Figures 4** and 5.

2. Propagation buckling of pipe-in-pipe systems

2.1 Introduction

Pipe-in Pipe (PIP) systems are extensively being used in the design of high pressure and high temperature (HP/HT) flowlines due to their outstanding thermal insulation. A typical PIP system consists of concentric inner and outer pipes, bulk heads and centralizers. The inner pipe (flowline) conveys the production fluids and the outer pipe (carrier pipe) protects the system from external pressure and mechanical damage. The two pipes are isolated by centralizers at joints and connected through bulkheads at the ends of the pipeline. The annulus (space between the tubes) is either empty or filled with non-structural insulation material such as foam or water [21].

PIP systems are normally divided into two categories, namely, compliant and non-compliant systems. In a compliant system, the inner pipe and the carrier pipe are attached at close intervals; whereas both inner and carrier pipes are only connected through bulkheads at discrete locations in a non-compliant system. The relative movement between the inner and outer pipes is arrested in a compliant system while the two pipes can move relative to each other in a non-compliant system. PIPs are exploited in subsea developments, where the carrier pipe is designed to resist high hydrostatic pressures (water depths up to 3000 m) and the inner pipe is designed to transmit hydrocarbons at temperatures as high as 180°C [22]. The HP/HT flow can cause global upheaval [23] or lateral buckling [24] in the system. Furthermore, the high hydrostatic pressure may trigger a local collapse, such as propagation buckling or buckle interaction [13–14, 25–29, 33], in the carrier (outer) pipe. Structural integrity of the PIP system under external pressure is an issue of concern, because the collapse of the carrier pipe may result in collapse of the inner pipe.

Despite the extensive investigations performed on integrity of single pipelines, to date, instabilities of PIPs have only been marginally addressed. Kyriakides [10] conducted a thorough experimental study on propagation buckling of steel PIPs with two-inch diameter carrier tubes with D_o/t_o values of 24.1, 21.1 and 16.7 and inner pipes of various D_i/t_i ratios ranging between 15 and 37. Kyriakides [10] observed two dominant modes of buckling. In the first mode the local collapse of the outer pipe led to simultaneous collapse of the inner pipe, whereas in the second mode the carrier pipe collapsed without affecting the inner pipe. Based on the experimental study and 3D finite element analyses, Kyriakides and Vogler [11] suggested an empirical formula for buckle propagation pressure of PIP, P_{p2}. Gong and Li [12] carried out a finite element study of propagation buckling of PIPs with carrier pipes having D_o/t_o values of 25, 20 and 15 and inner tubes having D_i/t_i of 15 and 20. Although both studies [11, 12] covered similar D_o/t_o range of the carrier pipe, two different empirical expressions were suggested.

2.2 Analytical solution of propagation pressure of pipe-in-pipe systems

Numerous analytical solutions have been suggested to estimate the propagation pressure of a single pipe. Unlike propagation pressure, the initiation pressure is very sensitive to initial imperfection such as local dents or ovalizations. The propagation pressure is related to plastic properties of the pipe and is only a fraction of the buckle initiation pressure. Both buckle initiation pressure and buckle propagation pressure are related to the diameter to wall-thickness ratio of the pipe, however previous studies suggest that there is no evident relationship between the two [2, 3]. The simplest propagation pressure model was established by Palmer and Martin [4], which only considered the initial and final configurations of the cross-section of the pipe. **Figure 6** shows the four plastic hinges developed in the pipe at different stages of propagation buckling on subsea pipelines and pipe-in-pipe systems.

By adopting plane strain analogy, Kyriakides and Vogler [11] proposed the following expression for the propagation pressure of the PIP system. Their formulation accounts for development of four plastic hinges in each of the carrier and the inner pipes (**Figure 6d-f**).

$$\hat{P}_{p2} = \frac{2\pi}{\sqrt{3}}\sigma_{Yo}\left(\frac{t_o}{D_o}\right)^2\left[1+\frac{\sigma_{Yi}}{\sigma_{Yo}}\left(\frac{t_i}{t_o}\right)^2\right] \tag{10}$$

where subscripts o and i denote the outer pipe and inner pipe, respectively.

The analytical lower bound solution to propagation buckling of a single pipe given by (Eq. (8)), can be extended to the pipe-in-pipe systems by accounting for the membrane and flexural effects of the outer and the inner pipes:

$$W_{ex} = W_{in(f)} + W_{in(m)} \tag{11}$$

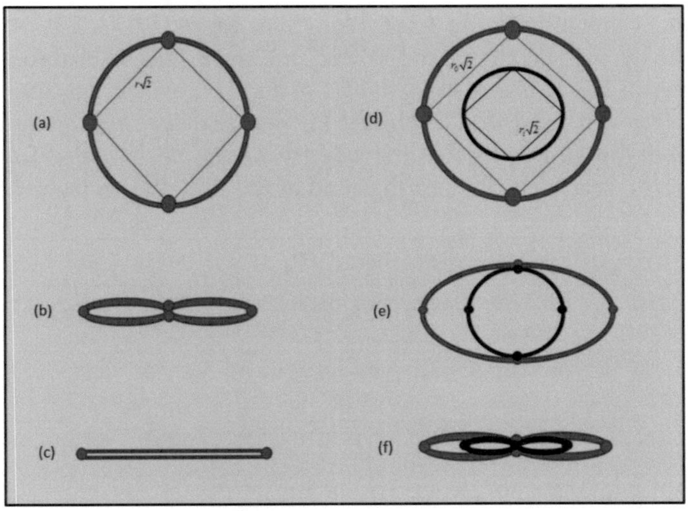

Figure 6.
A schematic of deformation stages in propagation buckling of a single pipe (stages a–c) and a pipe-in-pipe system (stages d–f).

where W_{ex} is the external work done by the hydrostatic pressure and W_{in} is the internal work due to the circumferential flexure, f and membrane effects, m. Based on the experimental observations from the hyperbaric chamber, the initially circular cross-section of the outer pipe (**Figure 6d**) has found to deform into the shape shown in (**Figure 6e**). Further increase in the external pressure causes the pipe-in-pipe system to eventually deform into the dog-bone shape (**Figure 6f**). Thus (Eq. (11)) can be written as:

$$\overline{P}_{p2}(\Delta A) = 3\pi(m_{po} + m_{pi}) + \overline{P}_{p2}(r_o.\Delta l_o + r_i.\Delta l_i) \tag{12}$$

where ΔA is the change in the cross-section area, Δl is the change in the circumferential length and mp is the plastic moment. These are given by:

$$\Delta A = \pi r_o^2 \tag{13}$$

$$\Delta l_o = \left(2\pi - 4\sqrt{2}\right) r_0; \quad \Delta l_i = \left(2\pi - 4\sqrt{2}\right) r_i \tag{14}$$

$$m_{po} = \sigma_{Yo}\frac{t_o^2}{4}; \quad m_{pi} = \sigma_{Yi}\frac{t_i^2}{4} \tag{15}$$

where the subscript "o" denotes the outer pipe, and "i" represents the inner pipe. Substituting Eqs. (13)–(15) into (12), the propagation pressure, $\overline{P}_{p2}$, of the PIP system is obtained as:

$$\overline{P}_{p2} = \left[\frac{3\pi\sigma_{Yo}}{2.515}\left(\frac{t_o}{D_o}^2\right)\right]\left[1 + \frac{\sigma_{Yi}}{\sigma_{Yo}}\left(\frac{t_i}{t_o}\right)^2\right]\left[\frac{1}{1-(D_i/2D_o)^2}\right] \tag{16}$$

When $D_i = t_i = 0$, Eq. (16) yields the propagation pressure of a single pipe given by Eq. (8). Unlike Eq. (10), Eq. (16) accounts for the effect of D_i/D_o as well as that of t_i/t_o and σ_{Yi}/σ_{Yo}.

2.3 Experiments on propagation buckling of pipe-in-pipe system

The experimental protocol is comprised of end-sealing concentric PIP systems with parameters shown in **Table 3** and a length of 1.6 m ($L/D > 20$), being pressurized inside the hyperbaric chamber depicted in **Figure 3a**. The chamber has an inner-diameter of 173 mm and a length of 4 m and is rated for working pressure of 20 MPa (2000 m water depth). The intact PIP was sealed at both ends by gluing on thick aluminum discs ensuring that the inner was completely sealed from the outer pipe. Two valves were connected to each end of the PIP, one on the carrier pipe and

ID	Carrier pipe	Inner pipe	D_o/t_o	D_i/t_i	D_i/D_o	t_i/t_o	E (MPa)	$\frac{E'}{E}$ (%)	σ_{Yo} (MPa)	$\frac{\sigma_{Yi}}{\sigma_{Yo}}$
PIP-1	OD = 80, t = 2	OD = 40, t = 1.6	40.0	25.0	0.50	0.80	69,000	1.01	169	0.93
PIP-2	OD = 60, t = 2	OD = 40, t = 1.6	30.0	25.0	0.75	0.80	69,000	0.97	139	1.12
PIP-3	OD = 80, t = 3	OD = 40, t = 1.6	26.7	25.0	0.50	0.53	69,000	1.02	209	0.75

Note: All dimensions are in millimeters; OD = *outer diameter;* t = *thickness.*

Table 3.
Properties of PIPs.

Figure 7.
The PIP sample inside the hyperbaric chamber and fittings.

the other on the inner pipe. One valve was used for bleeding the pipe while filling it with water. The second valve was utilized to vent the carrier and inner pipes, as well as to collect water from the inner pipe and the cavity between the inner and outer pipes during the buckle propagation (through the red and black hoses shown in **Figure 7**). A volume-controlled pressurization with a high pressure pump (shown in **Figure 3a**) was used and the pressure was increased until collapse of the system due to external pressure occurred under quasi-static steady-state conditions. By maintaining a low rate of pumping, the chamber pressure was stabilized at propagation pressure P_{p2}, with buckle longitudinally propagating along the PIP sample accompanied by water flow from the vents. The change in volume of the system (ΔV) during the test was calculated by measuring the weight of water being discharged from the inner pipe and the cavity between the pipes separately using digital weighing scales shown in **Figure 3a**. Control tests using a single pipe (outer pipe) were conducted first.

Figures 8–10 present the experimental results of the buckle propagation response of PIPs. The pressure inside the chamber is plotted against the normalized change in volume of the carrier pipe (60 × 2 mm) of PIP-2 in **Figure 8a**. The chamber is gradually pressurized until the initiation pressure P_I is reached at which a section of the pipe collapses resulting in a drastic drop in the chamber's pressure. The pressure is then maintained at the propagation pressure P_p with the dog-bone buckle shape longitudinally propagating along the length of the pipe. The buckle propagation response of the PIP-2 system is shown in **Figure 8b**. The change in pressure of the system is plotted against the normalized change in volume of the inner pipe (40 × 1.6 mm) and the outer pipe (60 × 2 mm) (the space between the two pipes). Buckle is initiated first (P_{I2}) on the outer pipe, then the energy is released through ovalization of the outer pipe, until the outer pipe touches the inner pipe. Buckle initiation pressures P_I and PI_2 have been shown to be closely related to geometric imperfections in shapes of dents or ovality of the outer pipe [25, 30]. Since the main focus of the present study is only on the buckle propagation pressures, the parameters affecting the buckle initiation pressure are not discussed herein. Following the contact between the carrier pipe and the inner pipes of PIP-2, the inner pipe collapses and the buckle propagates longitudinally as long as the pressure is maintained at P_{p2}. When the stiff end-caps fall within the vicinity of the buckle transition zone, a higher pressure is required to perpetuate the buckle which corresponds to the stiffening part of PIP-2 response shown in **Figure 8b**.

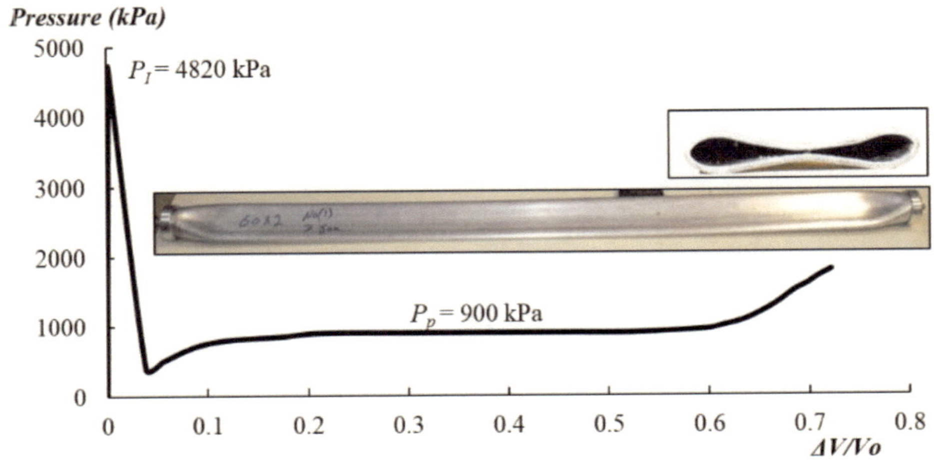

(a)

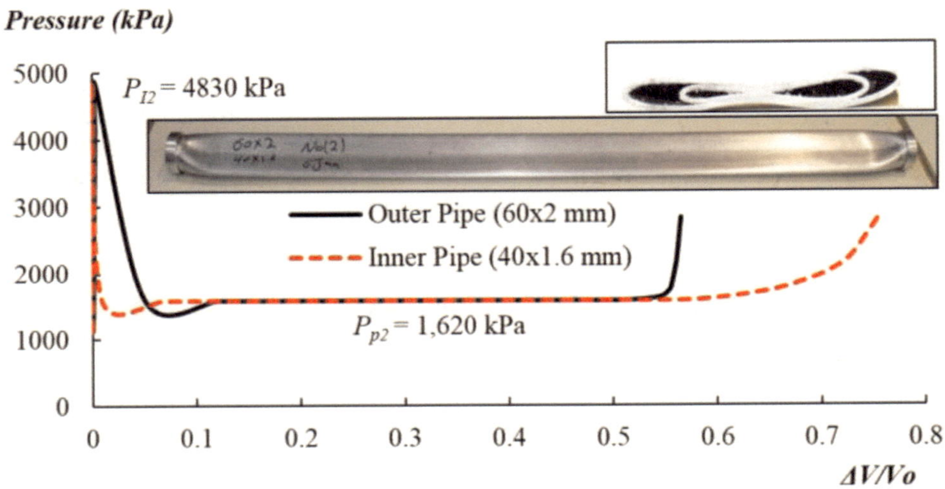

(b)

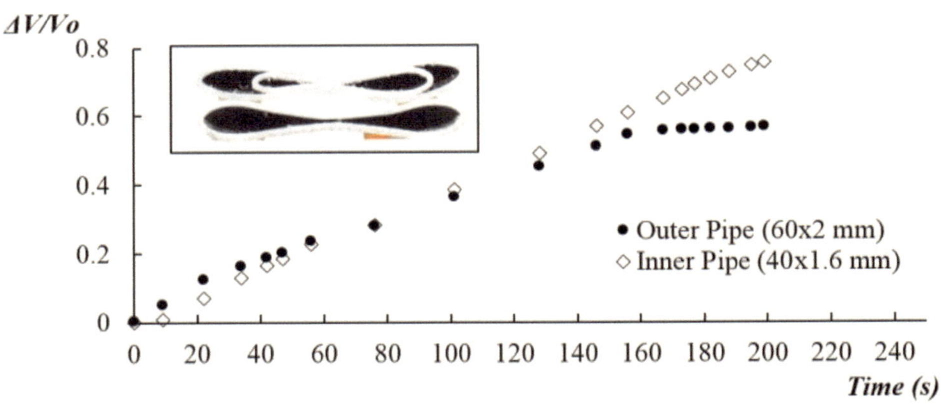

(c)

Figure 8.
Buckle propagation response inside the hyperbaric chamber: (a) pressure versus normalized change of volume of the 60 × 2 mm carrier pipe, (b) pressure versus normalized change of volume of PIP-2 and (c) normalized volume versus time of PIP-2.

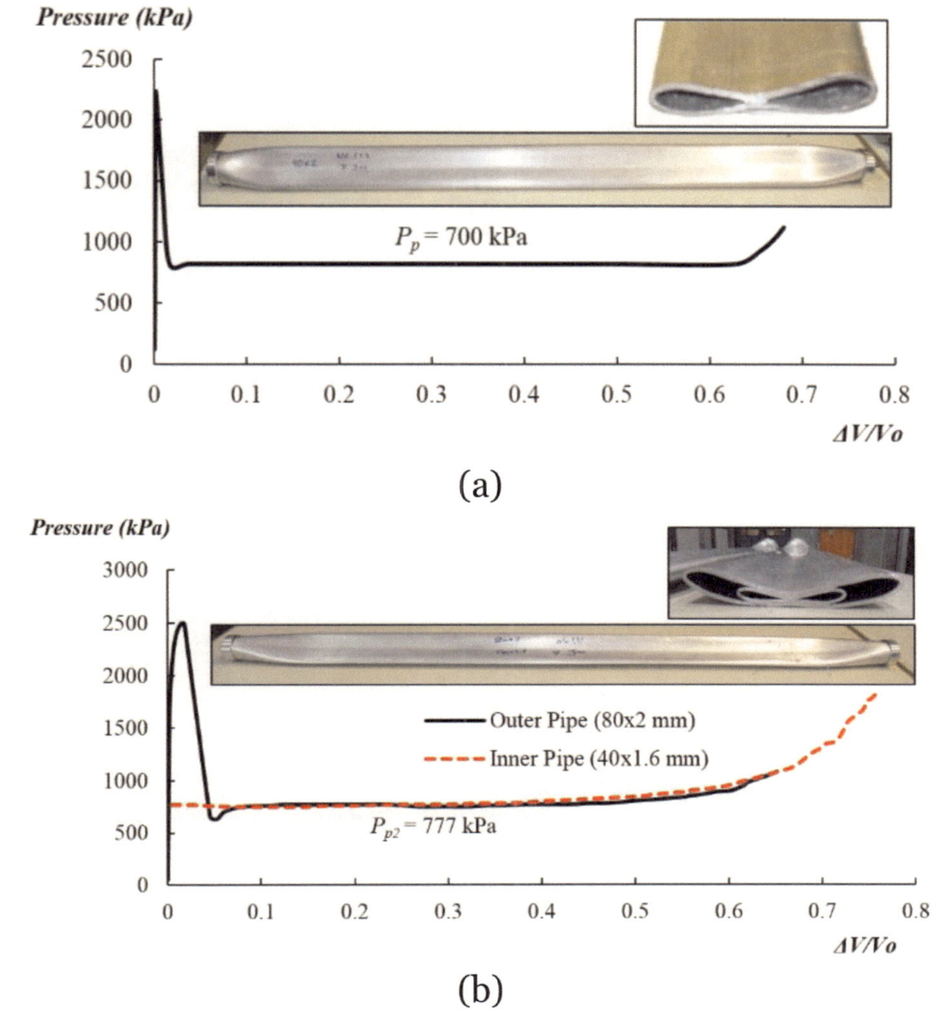

Figure 9.
Buckle propagation response inside the hyperbaric chamber: (a) pressure versus normalized change of volume of the 80 × 2 mm carrier pipe, (b) pressure versus normalized change of volume of PIP-1.

A dog-bone buckle shape similar to that observed in buckle propagation of the carrier pipe (**Figure 8a**) was detected in the PIP-2 chamber test (**Figure 8b**). Change in volume of the outer and inner pipes are plotted against the test time in **Figure 8c**. The time-history shows a higher initial discharge from the outer pipe than the inner pipe. However, after the outer pipe touches the inner pipe at $\Delta V/V_o = 0.1$ (shown in **Figure 8b**), discharge from the inner pipe is triggered and at $\Delta V/V_o > 0.2$ (shown in **Figure 8c**) the discharge rate of the inner pipe exceeds that of the outer pipe. This ascertains that the collapse of the outer pipe is rapidly transferred to the inner pipe and is then followed by the longitudinal propagation of the buckle in both carrier and inner pipes. The rate of discharge in the carrier pipe and inner pipe gradually decays as time lapses, which is due to the introduction of the end-caps in the buckle zone.

The hyperbaric chamber propagation buckling results of the 80 × 2 mm carrier pipe and the PIP-1 system are shown in **Figure 9**. A small dent was imposed to the carrier pipe in the single-pipe test, which explains the lower buckle initiation pressure of the carrier pipe compared to that of PIP-2. As shown in **Figure 9b**, following the collapse of the carrier pipe the pressure inside the chamber drops drastically

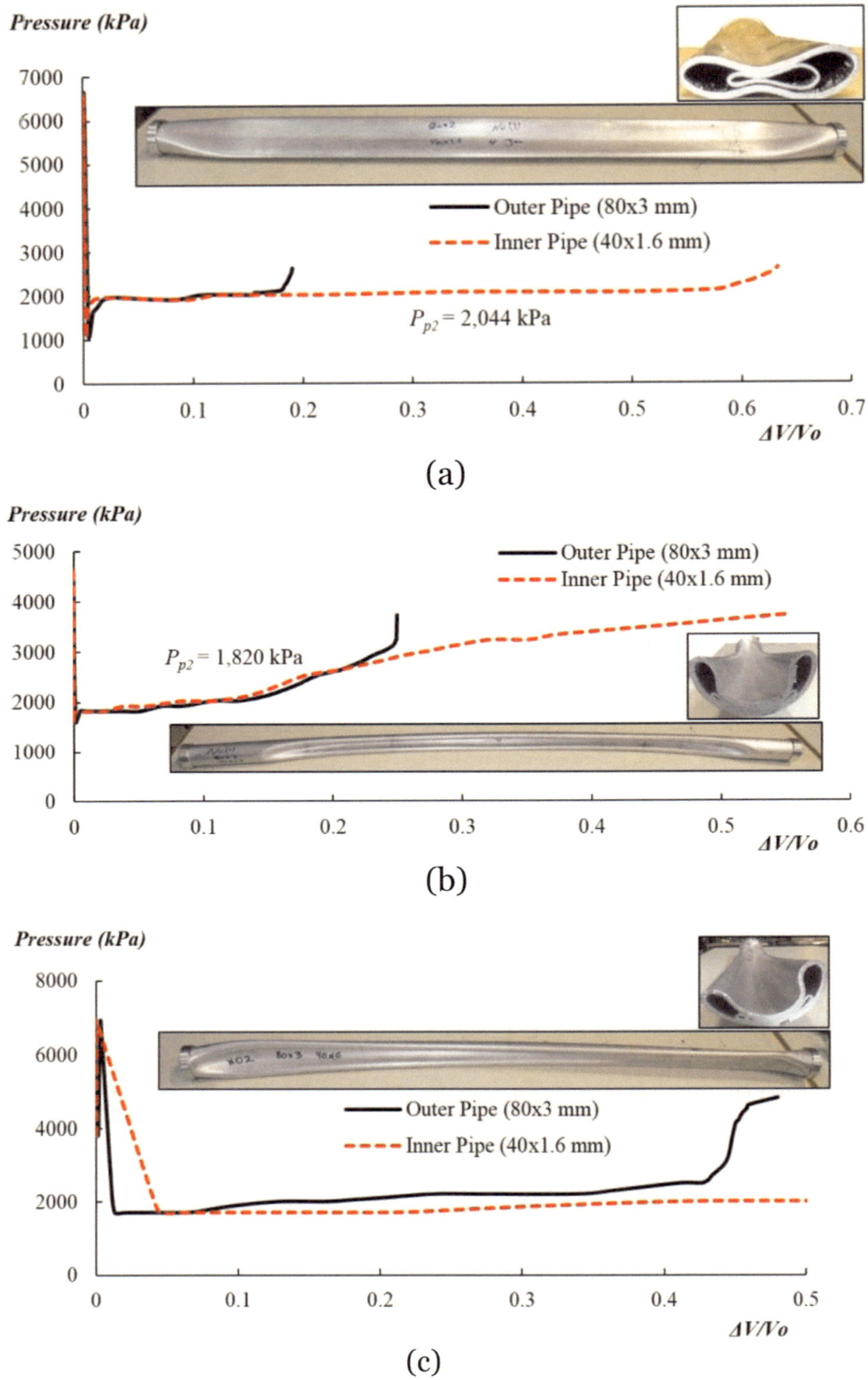

Figure 10.
Buckle propagation response inside the hyperbaric chamber: (a) buckle propagation response of the PIP-3 (80 × 3 mm-40 × 1.6 mm) with dog-bone buckle shape, (b) buckle propagation response of PIP-3 with confined buckle shape and (c) buckle propagation response of PIP-3 showing interaction between dog-bone and confined buckle shape.

until the carrier pipe and inner pipe come into contact. Subsequently, a dog-bone buckle shape propagates along the PIP while the pressure is maintained at P_{p2}. Hyperbaric chamber tests of PIP-1 and PIP-2 were repeated twice each and no significant disparities were observed in the results.

Results of the PIP-3 with D_o/t_o = 26.7 from three hyperbaric chamber tests are depicted in **Figure 10**. Unlike the responses of PIP-1 and PIP-2, three distinctive modes of buckling were observed in PIP-3, namely: (1) the dog-bone buckle shape (flat-mode) shown in **Figure 10a**, (2) the confined buckle shape (U-mode) shown in **Figure 10b** and (3) a combination of dog-bone and U-shaped buckle shown in **Figure 10c**. The dog-bone mode of buckling is similar to the responses observed in PIPs with high D_o/t_o values (PIP-1 and PIP-2). In this mode of failure, PIP-3 remains straight after failure and a flat mode of buckling propagates through its length; however the deformed shape of the inner pipe is not symmetric in the cross-section (shown in **Figure 10a**). In the second hyperbaric chamber test of PIP-3 shown in **Figure 10b**, a confined buckle shape is observed. The confined buckle mode is propagated along the length of the PIP while the pressure in the chamber is escalated followed by rapid discharge of water from the outer and inner pipes. It is worth mentioning that this U-shape buckling mode was previously observed in confined-buckling tests of steel and aluminum tubes reported by [31, 32]. Stephan et al. [32] performed an experimental investigation on the collapse of 3 m long aluminum pipes, inserted inside a 2 m long confining steel pipe. They observed a flat mode (dog-bone buckle shape) in the unconfined section of the aluminum pipe and a U-mode buckle shape within the confined section. Their experiments showed that within the studied range ($16 < D/t < 48$), the confined buckle shape consistently propagated at higher pressure compared to the dog-bone buckle shape. However the comparison between **Figure 10a** and **Figure 10b** shows that in PIP-3, the U-shape buckling propagation (P_{p2} = 1820 kPa) is initiated at a slightly lower pressure than the propagation pressure of the dog-bone buckle shape (P_{p2} = 2044 kPa). In the third test, PIP-3 showed a dog-bone failure mode that had flipped into a U-mode shape. The average P_{p2} results from all the hyperbaric chamber tests are given in **Table 4**.

2.4 Finite element analysis on propagation buckling of pipe-in-pipe systems

Finite element simulation of 1.6 m long samples of PIPs used in the hyperbaric chamber tests were conducted using ANSYS 16.1 [20]. Thin 4-node shell elements (181) were used to model the carrier pipe and the inner pipe. The contact between the inner and outer pipes, and in between the inner surfaces of the inner pipe were modeled using non-linear frictionless contact and target elements (174 and 170).

	Hyperbaric chamber		Analytical		FE
ID	**P_p (kPa)**	**P_{p2} (kPa)**	$\frac{\hat{P}_{p2}}{P_{p2}}$	$\frac{\widetilde{P}_{p2}}{P_{p2}}$	$\frac{P_{p2}^{FE}}{P_{p2}}$
PIP-1	700	780	0.78	0.86	1.28
PIP-2	900	1620	0.59	0.69	0.86
PIP-3	1400	2020*	0.64	0.66	0.96

Corresponds to dog-bone buckle shape shown in Figure 2.5a.

Table 4.
Comparison between hyperbaric chamber, analytical and FE results.

Symmetry is used and only one half of the pipeline is modeled. The mesh uses shell elements with seven integration points along the wall-thickness. To better facilitate the nonlinear analysis, a small ovalization ratio Ω (Eq. (9)) equal to 0.5% was introduced at mid-length on the carrier pipe in the FE model.

The nodes at either end of the PIPs were restrained from translation in all directions. A von Mises elastoplastic (bi-linear) material definition with isotropic hardening was adopted. The modulus of elasticity (E) and tangent modulus ($E/$) used in the FE models are also shown in **Table 3** and are based on the stress-strain curves obtained from the tensile longitudinal coupons taken from the pipe wall shown in **Figure 11a**. The yield stresses used in the FE models based on the stress-strain curves and are presented in **Table 3** as σ_{Yo} and σ_{Yi} for the outer pipe and inner pipe respectively. The FE predictions of the propagation pressure of PIP-2 and PIP-3 depicted in **Table 4** represent 86 and 96%, respectively, of the experimental results. However the propagation pressure obtained from the FE analysis overestimates the experimental results for PIP-1.

The pressure response and the deformed shape of PIP-1 from the FE analyses are shown in **Figure 11b**. The pressure is plotted against the normalized ovalization of the carrier and inner pipes ($\Delta D/D$). By increasing the hydrostatic pressure, the carrier pipe of PIP-1 in **Figure 11b** gradually deforms from the intact shape (I) into a deformed shape (II). At this stage the outer and inner pipes come into contact and a small deformation is observed in the inner pipe. The local collapse in the inner pipe is arrested which is followed by a slight increase in the pressure. The collapse is then propagated in the outer pipe until detained by the end-caps as depicted in the deformed shape (III). While the buckle approaches the endcaps, a higher pressure is required to maintain the volume inside the hyperbaric chamber. This increase, however, causes the inner pipe to collapse (IV). This mode of buckling in which the collapse propagates over the inner pipe was reported in [10, 12] to occur in a PIP system where the inner pipe is stiffer (has larger thickness and yield stress) than the outer pipe. However we observed this buckling mode in PIP-1, in which the inner pipe is softer than the outer pipe.

2.5 Empirical expressions for propagation buckling of PIPs with thin and moderately thin carrier pipes

A comprehensive parametric study is conducted using the validated FE model to find the buckle propagation pressures of PIP systems with various wall thickness t_i/t_o, diameter D_i/D_o, and the material yield stress σ_{Yi}/σ_{Yo} ratios. Prior to reviewing results of the parametric dependency of propagation buckling of PIPs, it is worth discussing the buckling modes observed in the FE simulations. The FE analyses showed two dominant modes of failure under external pressure in the studied PIPs. In a thin PIP (D_o/t_o of 40) shown in **Figure 12**, and with a thick ness ratio of t_i/t_o = 0.6 and identical outer and inner pipes, mode A is observed. In mode A, by increasing the external pressure, the carrier pipe collapses and gradually deforms from the undeformed shape (I) into the deformed shape (II). Then, the outer and inner pipes touch. The touchdown point corresponds to (II) in **Figure 12**. Then, the pressure needs to get larger so that the collapse propagates along both outer and inner pipes shown in stages (III) to (IV).

Figure 13 shows the pressure response and the deformed shape of a moderately thin PIP with D_o/t_o of 30 and t_i/t_o of 0.8. The outer and inner pipes are identical. Following the initiation of collapse in the outer pipe, the pressure in the system is dropped and the buckle is propagated in the carrier pipe as shown in deformed shapes of II and III in **Figure 13**. At (III) the collapse has reached the end caps, therefore, a higher pressure is required to perpetuate the collapse in the outer pipe. However the increase in pressure causes a collapse in the inner pipe at the pressure

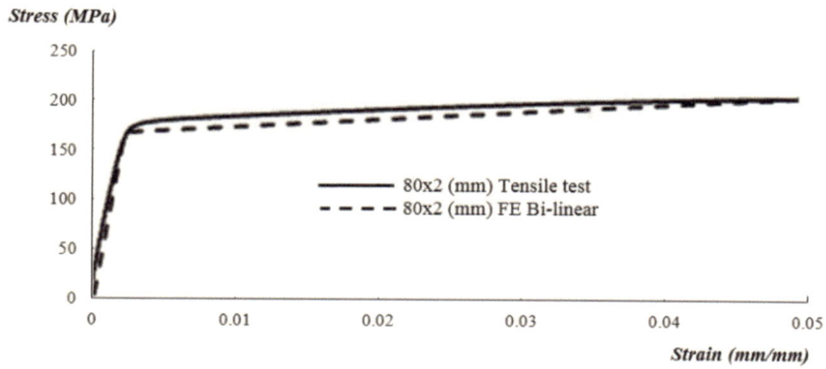

(a)

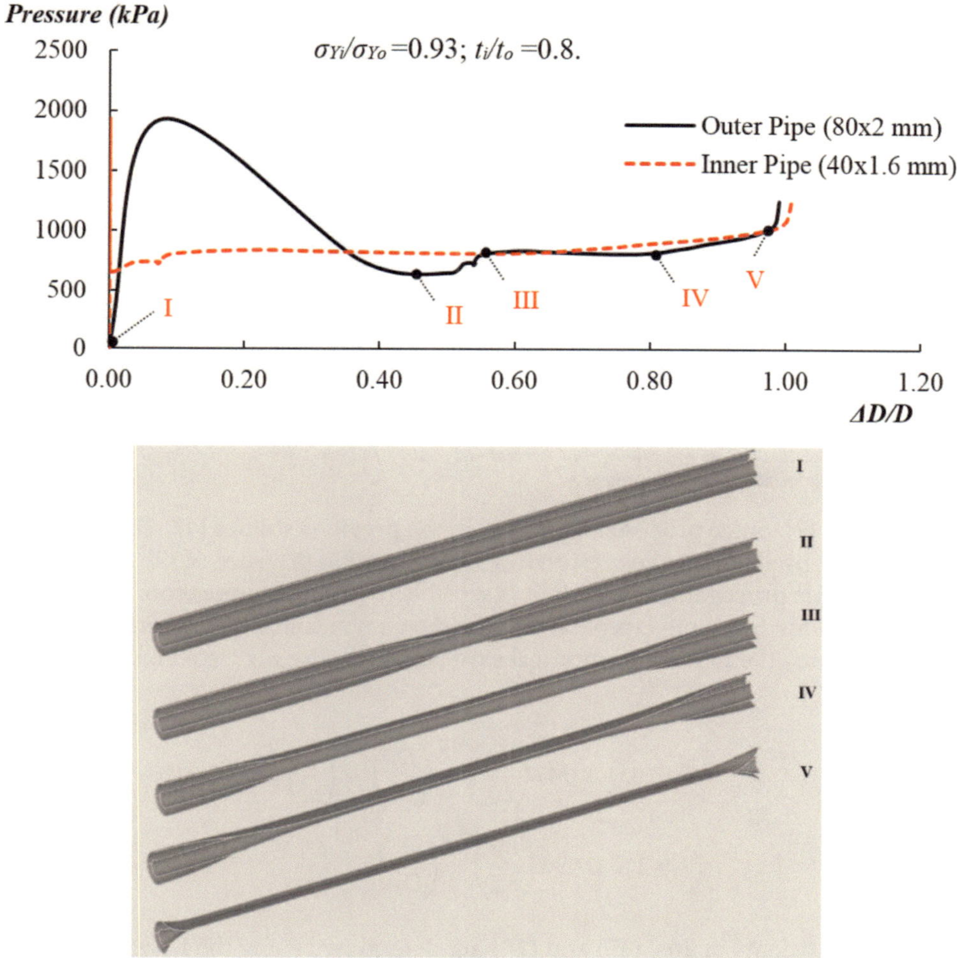

(b)

Figure 11.
(a) Experimental and FE stress–strain curves; (b) FE results showing pressure against normalized ovality and corresponding PIP-1 deformed shapes.

level (IV) and initiates a buckle which is propagated through the length (V). This buckle propagation mode is referred to as mode *B* in this study.

The parametric study ascertained the dependency of the propagation pressure of the PIP systems on geometric and material parameters of the outer and inner pipes. Moreover, current FE results proved that the buckle propagation modes of PIPs with large D_o/t_o ratios are not essentially similar to mode *A* predicted in previous

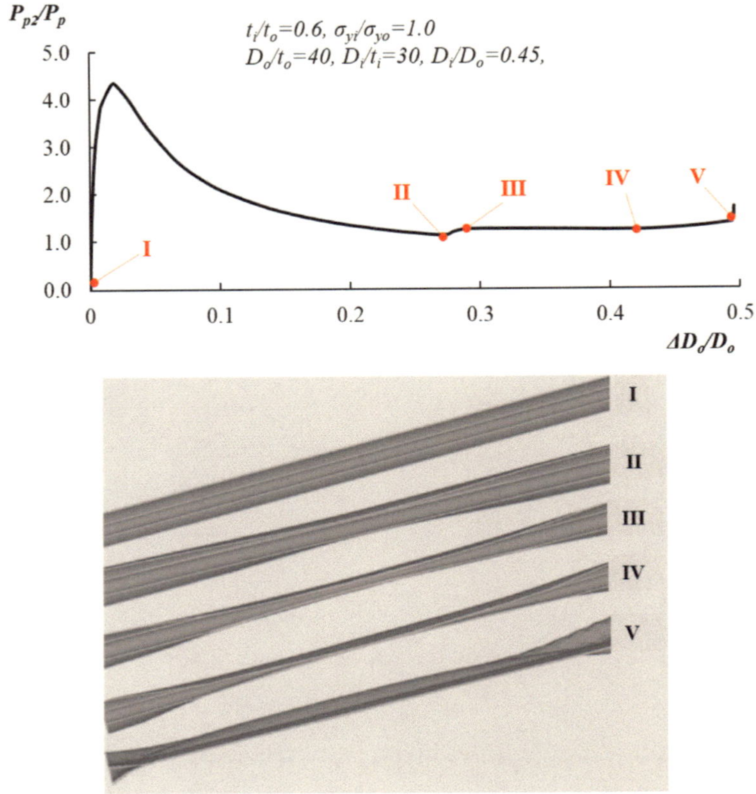

Figure 12.
Finite element results showing pressure against normalized ovality and corresponding deformed shapes of PIP system exhibiting buckle propagation mode A.

studies [11, 12]. Since proposed equations in the previous studies [11, 12] are incapable of predicting proper estimates of propagation pressure of PIPs that exhibit buckle propagation mode *B*, it is sensible to propose expressions for buckle propagation modes *A* and *B* separately. Based on a non-linear square fits of all the FE data points, the following empirical expressions can be suggested for Modes *A* and *B*.

$$\frac{P_{p2}}{P_p} = 1 + 1.047\left(\frac{\sigma_{yi}}{\sigma_{yo}}\right)^{0.2}\left(\frac{D_i}{D_o}\right)^{0.4}\left(\frac{t_i}{t_o}\right)^{2.4} \tag{17}$$

$$\frac{P_{p2}}{P_p} = 1 + 0.596\left(\frac{\sigma_{yi}}{\sigma_{yo}}\right)^{0.2}\left(\frac{D_i}{D_o}\right)^{-0.8}\left(\frac{t_i}{t_o}\right)^{2.4} \tag{18}$$

The coefficients in Eqs. (17) and (18) are determined using the Leven-berg-Marquardt algorithm and correspond to correlation factors (R^2) of 0.9827 and 0.9860 respectively. Comparison between the FE results and the proposed expressions are shown in **Figures 14a,b** for buckle propagation modes *A* and *B* respectively. The maximum differences between FE results and empirical expressions are less than 6.0%.

2.6 Empirical expressions for propagation buckling of PIPs with thick and moderately thick carrier pipes

In PIP systems with thin and moderately thin carrier pipes, expressions (Eqs. (17) and (18)) derived in Section 2.5 can be used to predict the propagation

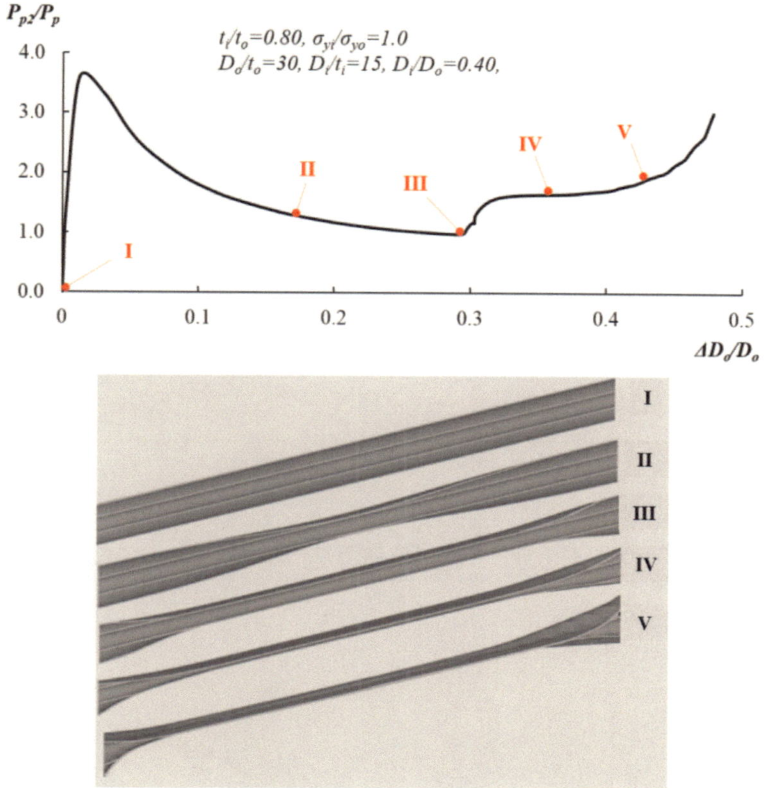

Figure 13.
Finite element results showing pressure against normalized ovality and corresponding deformed shapes of PIP system exhibiting buckle propagation mode B.

pressures. A total of 254 data points were collected from the raw data reported in [11, 12], and the current FE results for PIPs with D_o/t_o = 26.67. These data were used to propose an expression to predict the propagation pressure of PIPs with thick and moderately thick carrier pipes. Using the Levenberg-Marquardt algorithm of non-linear least squares the following expression was derived for the propagation pressure, P_{p2}, of PIPs with D_o/t_o < 27.

$$\frac{P_{p2}}{P_p} = 1 + 0.803\left(\frac{\sigma_{yi}}{\sigma_{yo}}\right)^{0.4}\left(\frac{D_i}{D_o}\right)^{0.13}\left(\frac{t_i}{t_o}\right)^{1.8} \tag{19}$$

with multiple correlation factor (R^2) of the fit is 0.9781. Similar procedure is used to derive (Eq. (19)). The expression accounts for the interaction between non-dimensional variables. For sake of brevity, the procedure is not shown here. Finally, the current FE results, the FE results of [12] and experimental results of [11] are collated in **Figure 15**, and are plotted against the proposed expression (Eq. (19)). The plot forms a nice linear band. The results in **Figure 15** correspond to buckle propagation mode *A*.

2.7 Empirical expression for collapse pressure Pci of PIPs

The hyperbaric chamber results disused in the previous section suggest that, the collapse pressure of the inner pipe of the PIP system, (P_{ci}), is a function of geometric and material parameters of both inner and outer pipes. A comprehensive parametric study carried out herein ascertained the dependency of the collapse pressure

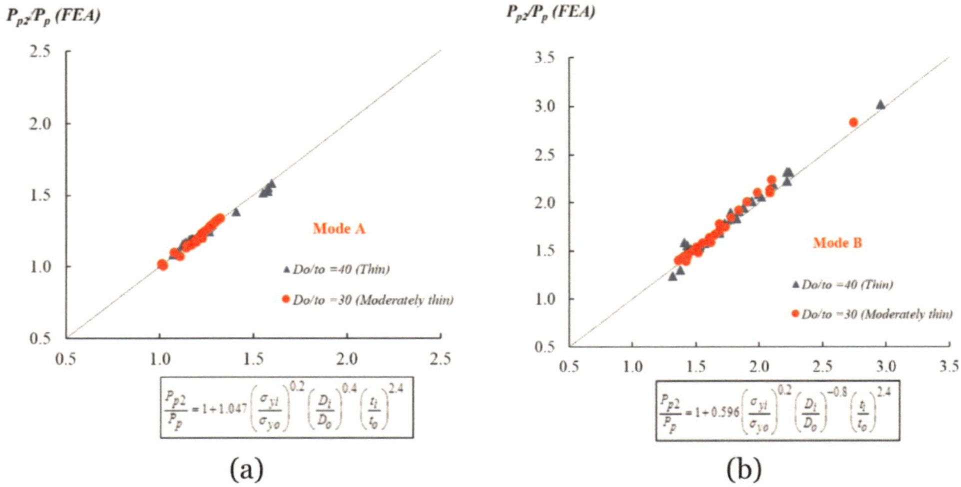

Figure 14.
Comparison between FE results and those predicted by Eqs. (17) and (18) of buckle propagation pressures of PIP with buckle propagation (a) mode A; and (b) mode B.

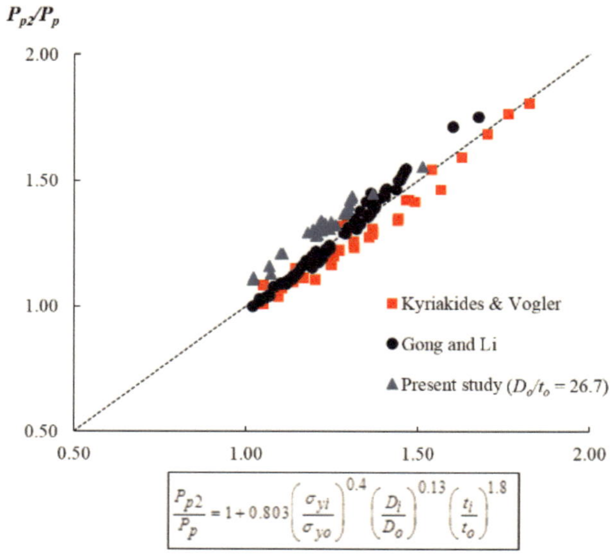

Figure 15.
Comparison between buckle propagation pressures of thick to moderately thick PIP systems from previous studies and current expression (all results correspond to the buckle propagation mode A).

P_{ci} of the PIP systems on geometric and material parameters of the outer and inner pipes. Based on the results of the parametric study and using non-linear square fits of sets of data taken from the FE results, the following normalized expression is derived for the collapse pressure of the inner pipe of PIPs.

$$\frac{P_{ci}}{Pcr} = 0.05\left(\frac{D_i}{D_o}\right)^{3.2}\left(\frac{t_i}{t_o}\right)^{-1.88}\left(\frac{D_i}{t_i}\right)^{0.68}\left(\frac{\sigma_{yi}}{\sigma_{yo}}\right)^{-0.6}\left(\frac{E'_i}{E'_o}\right)^{-0.3} \quad (20)$$

The coefficient (0.05) in Eq. (20) is determined using the Levenberg-Marquardt algorithm with a correlation factor (R^2) of 0.9882. Comparison between the FE results and the proposed expression (Eq. (20)) is depicted in **Figure 16** for the

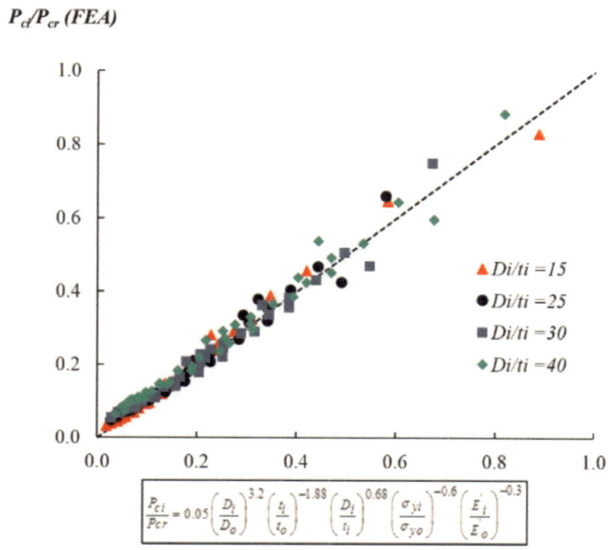

Figure 16.
Comparison between FE results and those predicted by Eq. (20).

	P_{ci}/P_{cr} (Eq. (20))	P_{ci}/P_{cr} (Exp.)	Difference (%)
PIP-1	0.173	0.166	4.05
PIP-2	0.077	0.077	0.00
PIP-3	0.188	0.184	2.13

Table 5.
Comparison between empirical and experimental collapse pressures.

studied range of D_i/t_i. The maximum difference between FE results and empirical expression (Eq. (20)) is less than 6.0%.

The normalized collapse pressures obtained from the proposed empirical expression (Eq. (20)) and those acquired from the hyperbaric chamber for the tested PIPs are represented in **Table 5**. The differences are less than 5%. As represented in the last column of **Table 5**, the empirical expression predicts the experimental results with good accuracy.

3. Conclusion

Buckling propagation mechanisms of subsea single-walled pipelines and pipe-in-pipe (PIP) systems under external pressure in quasi-static steady-state conditions were investigated using 2D analytical solutions, hyperbaric chamber and 3D FE analyses considering non-linear material and geometric behavior. In general, reasonable agreement is obtained between analytical, numerical and experimental results. The modified analytical solution suggested in this chapter accounts for the D_i/D_o ratio and provides more accurate predictions of the propagation buckling pressure of PIPs compared to the previous analytical equations. Confined buckling and flip-flop buckling modes were discovered in the hyperbaric chamber test of PIP-3 9 (**Table 3**). Nonlinear finite element analyses were conducted and verified against the hyperbaric chamber tests. The FE models provided valuable information about the buckling modes and progress in the carrier and inner pipes.

The comprehensive FE study suggested the existence of two major buckle modes in PIPs with thin and moderately thin carrier pipes. In mode *A* the buckle propagated simultaneously in the outer and inner tubes and in mode *B* the buckle propagated in the outer pipe and the collapse in the inner pipe was delayed. For each buckling mode, a separate expression was proposed, (Eqs. (17) and (18)). Based on the combined data from previous studies and current FE results, a more accurate empirical expression (Eq. (19)) was proposed to predict the propagation pressure P_{P2} of PIPs with thick and moderately thick carrier pipes. Moreover, the collapse pressure of the inner pipe of the PIP (P_{ci}) system was formulized. The proposed expression was shown to be in good agreement with hyperbaric chamber test results.

Author details

Hassan Karampour* and Mahmoud Alrsai
Griffith School of Engineering and Built Environment, Griffith University, QLD, Australia

*Address all correspondence to: karampour@griffith.edu.au

References

[1] Kyriakides S, Corona E. Mechanics of Offshore Pipelines: Volume 1 Buckling and Collapse. Vol. 1. Oxford, UK: Elsevier; 2007

[2] Albermani F, Khalilpasha H, Karampour H. Propagation buckling in deep sub-sea pipelines. Engineering Structures. 2011;**33**(9):2547-2553

[3] Mesloh RE, Sorenson JE, Atterbury TJ. Buckling and offshore pipelines. Gas. 1973;**49**(7):40-43

[4] Palmer AC, Martin JH. Buckle propagation in submarine pipelines. Nature. 1975;**254**(5495):46

[5] Kyriakides S, Babcock CD. Experimental determination of the propagation pressure of circular pipes. Journal of Pressure Vessel Technology. 1981;**103**(4):328-336

[6] Kyriakides S, Netto TA. On the dynamics of propagating buckles in pipelines. International Journal of Solids and Structures. 2000;**37**:6843-6867

[7] Kamalarasa S, Calladine CR. Buckle propagation in submarine pipelines. International Journal of Mechanical Sciences. 1988;**30**(3–4):217-228

[8] Palmer AC, King RA. Subsea pipeline engineering. Oklahoma, USA: PennWell Books; 2004

[9] Xue J, Fatt MH. Buckling of a non-uniform, long cylindrical shell subjected to external hydrostatic pressure. Engineering Structures. 2002;**24**(8): 1027-1034

[10] Kyriakides S. Buckle propagation in pipe-in-pipe systems: Part I. Experiments. International Journal of Solids and Structures. 2002;**39**(2): 351-366

[11] Kyriakides S, Vogler TJ. Buckle propagation in pipe-in-pipe systems: Part II. Analysis. International Journal of Solids and Structures. 2002;**39**(2): 367-392

[12] Gong S, Li G. Buckle propagation of pipe-in-pipe systems under external pressure. Engineering Structures. 2015; **84**:207-222

[13] Karampour H, Alrsai M, Albermani F, Guan H, Jeng DS. Propagation buckling in subsea pipe-in-pipe systems. Journal of Engineering Mechanics. 2017; **143**(9):04017113

[14] Alrsai M, Karampour H, Albermani F. Numerical study and parametric analysis of the propagation buckling behaviour of subsea pipe-in-pipe systems. Thin-Walled Structures. 2018; **125**:119-128

[15] Estefen SF, Netto TA, Pasqualino IP. Strength analyses of sandwich pipes for ultra deepwaters. Journal of Applied Mechanics. 2005;**72**(4):599-608

[16] Showkati H, Shahandeh R. Experiments on the buckling behavior of ring-stiffened pipelines under hydrostatic pressure. Journal of Engineering Mechanics. 2009;**136**(4): 464-471

[17] Timoshenko SP, Gere JM. Theory of Elastic Stability. NewYork, USA: McGraw-Hill; 1961

[18] Kyriakides S, Babcock CD. On the dynamics and the arrest of the propagating buckle in offshore pipelines. In: Offshore Technology Conference. 1979;(1):45-57

[19] Mesloh R, Johns TG, Sorenson JE. The propagating buckle. Boss. Proceedings. 1976;**76**:787–797

[20] ANSYS 16.1 Release, A. I. 275 Technology Drive, Canonsburg, PA 15317

[21] Bai Y, Bai Q, editors. Subsea Pipelines and Risers. Oxford, UK: Elsevier; 2005

[22] Jukes P, Eltaher A, Sun J, Harrison G. Extra high-pressure high-temperature (XHPHT) flowlines: Design considerations and challenges. In: Proceedings of the ASME 2009 28th International Conference on Ocean, Offshore and Arctic Engineering. American Society of Mechanical Engineers. 2009. pp. 469-478

[23] Wang Z, Chen Z, Liu H. Numerical study on upheaval buckling of pipe-in-pipe systems with full contact imperfections. Engineering Structures. 2015;**99**:264-271

[24] Vaz MA, Patel MH. Lateral buckling of bundled pipe systems. Marine Structures. 1999;**12**(1):21-40

[25] Karampour H, Albermani F. Experimental and numerical investigations of buckle interaction in subsea pipelines. Engineering Structures. 2014;**66**:81-88

[26] Karampour H, Albermani F. Buckle interaction in textured deep subsea pipelines. Ships and Offshore Structures. 2016;**11**(6):625-635

[27] Karampour H, Albermani F, Gross J. On lateral and upheaval buckling of subsea pipelines. Engineering Structures. 2013;**52**:317-330

[28] Karampour H, Albermani F, Major P. Interaction between lateral buckling and propagation buckling in textured deep subsea pipelines. In: Proceedings of the ASME 34th International Conference on Ocean, Offshore and Arctic Engineering. American Society of Mechanical Engineers; 2015. pp. V003T02A079-V003T02A079

[29] Karampour H, Albermani F, Veidt M. Buckle interaction in deep subsea pipelines. Thin-Walled Structures. 2013; **72**:113-120

[30] Zheng J, Palmer A, Brunning P, Gan CT. Indentation and external pressure on subsea single wall pipe and pipe-in-pipe. Ocean Engineering. 2014;**83**: 125-132

[31] Lee LH, Kyriakides S. On the arresting efficiency of slip-on buckle arrestors for offshore pipelines. International Journal of Mechanical Sciences. 2004;**46**(7):1035-1055

[32] Stephan P, Love C, Albermani F, Karampour H. Experimental study on confined buckle propagation. Advanced Steel Construction. 2016;**12**(1):44-54

[33] Alrsai M, Karampour H, Albermani F. On collapse of the inner pipe of a pipe-in-pipe system under external pressure. Engineering Structures. 2018; **172**:614-628

Chapter 9

The Influence of Water Quality on the Structural Development of Vessels: Smart Dimensioning Process

Sérgio António Neves Lousada, João Pedro Gouveia and Rui Alexandre Castanho

Abstract

In fact, it is pivotal the development, use, and management of the best and most suitable coatings to be supplied to vessels—especially those designed for long journeys; not only to increase their stability and safety but also to minimize their maintenance cost. In this regard, it should be also considering the function, the vessel typology and its routes, as well as the quality of the waters by which it will navigate. Topics that are critical to promote a better dimensioning process of vessels. Thus, the present chapter, *via* an extensive literature review articulated with practical approaches, aims to define relevant directions for vessels structural development processes regarding the water quality (sea or river waters), where they will outline their routes. Therefore, the study looks for a relationship between the vessels structural coating design process and the quality of the water where they navigate. Moreover, such a process not only will optimize/minimize the costs with the periodic maintenance of the vessels linings, but also to relate it with its routes—contributing to the revitalization of their structural dimensioning.

Keywords: dimensioning, vessels, water quality, management of maritime transportation navigation routes

1. Introduction

Nowadays more than ever has increased the need to preserve the environment, in that sense this chapter intends to make a slight contribution to this issue.

It is never too much to emphasize the importance of the sea, which occupies about 71% of the Earth's surface, has an enormous importance in life on Earth. Suffice it to say, that 70% of the oxygen produced on Earth is generated by the marine phytoplankton, being a crucial element for the existence of life. The sea has a fundamental and decisive importance in the climate that surrounds the Earth for example the cloud formation process, fundamentally based on seawater evaporation, then filtered, and transformed into freshwater. Data provided by USGS Water Science School [1].

Water is precious because life could not exist without it. Life was born has a result of water, from plants to animals to humans [2].

Another colossal factor inherent to the sea are the winds, necessary to the pollination of the plants, originated from the variation in temperature between the land and the sea. Moreover, water transportation has a huge impact in socio-economical life of humans, although in recent centuries has contributed less positively to the environment.

Furthermore, recently the sea and the oceans have a more important role energy wise, with a very promising future as an ecological renewable source. Additionally the sea hosts offshore wind farms and there's an increasing trend from several countries to use floating solar plants.

Maritime transportation has played an important role in international goods transportation. Large ships have a massive load capacity and consume large amounts of fossil fuels to operate high-energy consumption entails high pollutant emissions that have adverse impacts on the marine and atmospheric environment and on public health.

In order to minimize potential impacts, enhance environmental opportunities and provide an environmentally sustainable waterway system, it is necessary to identify, quantify and predict vessel effects and their potential ecological impacts [3].

According to Andria et al. (2008), monitoring and modeling studies are useful and suitable tools for assessing the environmental pollution. In this regard many studies have been made, however in order to mitigate this problem, it will require the implementation of more eco-friendly coatings and more resistant to seawater erosion.

Globally, coastal water quality is deteriorating due to anthropogenic influences [4].

Unfortunately, despite everything previously mentioned, human activities still negatively influence water quality and aquatic ecosystem functions, resulting in a decline of water quality, biodiversity, loss of critical habitats and an overall decrease in the quality life of all species. Although this effect is more meaningful in some regions, it affects the populations around the globe.

The overall objectives of the casuistic approached is to determine the extent to which the vessels have unfavorably contributed to the increase in pollution and to provide some guidelines regarding the structural design processes in order to mitigate the harmful consequences of the non-preservation of our planet, particularly for navigable waters.

2. State of the art

2.1 Maritime transportation

Maritime transportation plays an important role in the world merchandise trade and economics development. Most of the large volume cargo between countries like crude oil, iron ore, grain, and lumber are carried by ocean vessels [5].

According to UNCTAD (United Nations Conference on Trade and Development) Review of Maritime Transport [6, 7], Maritime transport is the backbone of international trade and a key engine driving globalization. Around 80% of global trade by volume and over 70% by value is carried by sea and is handled by ports worldwide. These shares are even higher in the case of most developing countries. Only to demonstrate the importance of maritime transport to humankind, specifically the sea born trade, here are some facts illustrated in **Figures 1, 2** and **3**.

2.2 Pollution caused by shipping activities

The main components of ship resistance consist of resistance due to wave resistance, pressure resistance, and frictional resistance. With the improvement of

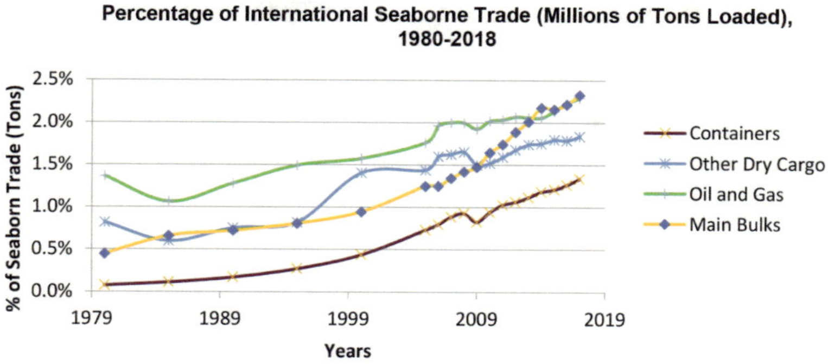

Figure 1.
Percentage variation through the years by category of international seaborne trade.

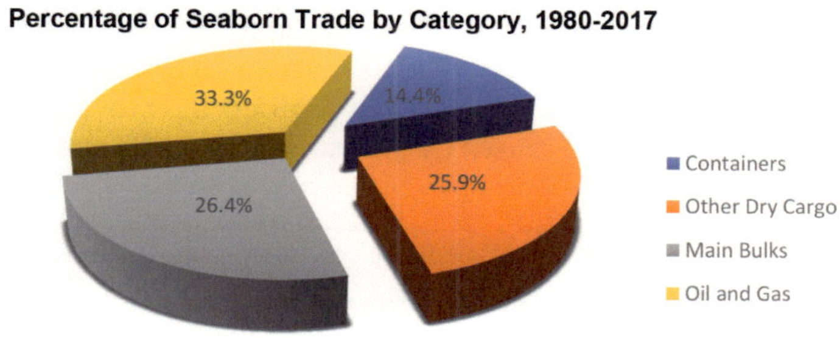

Figure 2.
Percentage of seaborne trade by category from 1980 to 2017.

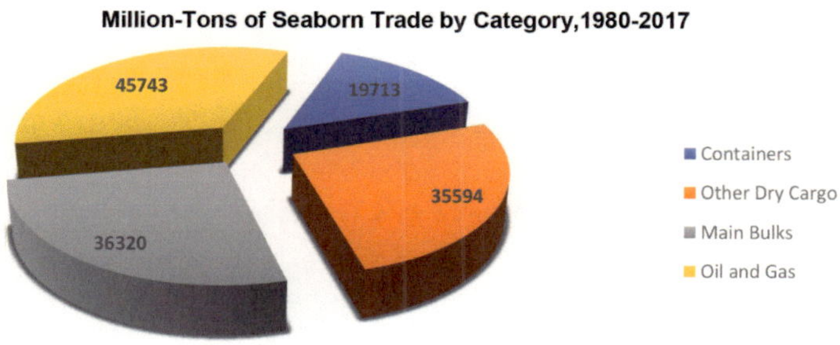

Figure 3.
Amount of seaborne trade in million-tons by category from 1980 to 2017.

hull form optimization techniques, the wave and the pressure components could be less than 20% of the total drag in most modern ships. Therefore, the advantage from the reduction of the remaining frictional drag would be enormous [8].

The CO_2 emission from international marine bunkers in 2014 was estimated 626.1 million tons. Considering the conversion ratio 3.17 between CO_2 emission and fuel consumption [9], this amounts to 197.5 million tons of fuel consumption, which corresponds to approximately 60 billion US$/year. Thus, 10% reduction of frictional drag would lead to saving of 4.7 billion US$/year.

A large number of vessel worldwide still uses non-eco-friendly paints. In a sea environment, paint erosion is inevitable. Therefore, these type of paints must be replaced sooner rather than later, in order to reduce its negative footprint in marine ecosystems. In the specific case of Madeira Island, there are two main ports, an industrial one, the Port of Caniçal and a mainly touristic one, the Port of Funchal. Besides that, there is a significant amount of fishing, touristic and recreational shipping activities in Madeira, which proves, the need to ascertain the type of ink most used, the periodicity with which the maintenance is made, how degraded the coatings are.

The vessels that least care about the esthetics of the ships, are those of industrial character as well as those for fishing activities. This seems harmless, but has serious environmental implications, on the one hand polluting the sea, another of the implications is that the degradation of paints causes a significant decrease in the speed of navigation and consequently an increase in the fuel consumption, thus polluting the air and increasing the already uncontrolled use of fossil fuels. Although the scenario is already bad, its worse than it seems, since the vessels that consume more fuel are the ones previously mentioned, for example the cargo ships transporting astronomical loads, would benefit and much of a more restrictive and periodic maintenance. In this regard, it matters to apply higher quality (in an ecological point of view) coatings, with less friction. This type of measure will allow lowering the consumption and increasing the efficiency. The above is explained is presented in **Table 1**.

According to consultant Det Norske Veritas, fuel costs constitute the largest expense of shipping companies and the drop in oil prices has brought some relief to the reduction in freight rates, caused by the excess of capacity and slowdown in

NSTM rating	Description	Increase in power at 15 kn
0	Hydraulically smooth surface	0.00%
0	Typical as applied antifouling coating	2.00%
10–20	Deteriorated coating or light slime	11.00%
30	Heavy slime	21.00%
40–60	Small calcareous fouling or weed	35.00%
70–80	Medium calcareous fouling	54.00%
90–100	Heavy calcareous fouling	86.00%

Table 1.
The effect of the roughness of the coatings on the power increment required for a speed of 15 knots.

Figure 4.
Typical pollution sources from vessels.

global growth. According to the same source, reducing the speed of a ship by 10% can lower fuel consumption by about 30%.

Figure 4 shows that although the fuels and paints are the most preponderant pollutants, there are many more originated by vessels.

Based on the foregoing it is of paramount importance to preserve and protect this source of life, the next chapter aims to offer a small contribute to improve and optimize the shipping industry, giving continuity to the advances already made.

3. New trends and directions for a smart dimensioning process

3.1 Smart design

Smart design - Smart design consist of a combination of techniques and tools which aid smart product design, including computational intelligence, SBD, design automation, e.g. In this case, by automating the simulation of product scenario using SBD and computational intelligence techniques, it can helps in the rapid testing and development of innovative design, which in turn translate to better products and revenue [10].

Some cruise ships companies are genuinely concerned with the negative impact caused by vessels, for this reason, it is beneficial that they have already managed to combat and reverse this trend. For instance Holland America ms Oosterdam *Ship* has done a very thorough job considering every aspect aboard their vessel and improvements for both energy efficiency and waste reduction. The implementation of Black water treatment system. In addition, they have reduced their 8 tons of waste generated each day by working with their supply chain, waste does not go overboard. Another significant measure was the incorporation of environmental officers on every cruise as well as staff eco-educated.

In an increasingly less utopian perspective, ships design should take into account the reduction of their own weight, as well as the use of materials that do not rust. Another key issue that has been addressed is the reduction of drag effect. It matters that coatings, fluctuation capacity and the aeronautical itself must be astronomically optimized. An additional key factor is the energy consumption of ships, the ship designer in this decade and in the future has the duty to select environmentally friendly engines and the ship itself has to be an autonomous power generation center. Robotics and current informatics will be the best means to increase productivity, maintenance and construction of ships, but this collaboration between man–machine has to be regulated in order not to cause unemployment, but rather help reducing the workload of employees. Lastly, but not least important, water and waste treatments should exponentially reduce their effect, aiming for ships to operate equivalently to Smart Cities, thus equipped with latest generation WWTP (Wastewater Treatment Plant) as well as SWTP (Solid Waste Treatment Plant).

3.2 Lightweight materials launched for shipbuilding

Sustainable solutions such as lightweight construction techniques and advanced materials are in demand, e.g. as the RAMSSES Project (Realization and Demonstration of Advanced Material Solutions for Sustainable and Efficient Ships) has the strategic objective to obtain recognition and an established role for advanced materials in the European Maritime Industry. Image credit: Evonik number of container ships in operation is constantly growing in response to the global volume of commercial trading. Ships equipped with the new hulls will be less expensive to operate relative to steel construction due to lower fuel demands and increased cargo capacity.

Fiber-reinforced composites do not rust, and their excellent resistance to seawater will translate into a reduced need to renew protective finishes and extended maintenance intervals.

Applying lightweight (primarily organic) material to high performance monohull sailing hulls/decks has resulted in a proportional increase in beam. Fully loaded container ships have a very broad "sail" area, so some amount of broadening of hull beam may be expected. Due to this structural modification, the navigational routes, ports and channels, which will be used by these vessels, will have to adapt quickly and at last resort to carry out the necessary construction works.

3.3 Super hydrophobic coatings

One of the super hydrophobic coating known is AIRCOAT (Air Induced friction Reduction Coating), works similarly to Salvinia leaves by creating a thin air layer that acts as a physical barrier between the water and the outside of the ship. This particular coating helps to reduce fuel oil consumption and gas emissions, as less energy will be required to move the ship forward, making transport more sustainable. The air barrier created by AIRCOAT will also help reduce the attachment of bacteria and algae that cause fouling. The corrosion of metals can produce a premature failure of metallic components, resulting in financial losses, environmental contamination, as well as injury or death [11, 12]. There's a much higher number of super hydrophobic coatings in the market, never the less the accession in the naval industry is still very distant from the desirable.

3.4 Energetically autonomous ships

Research on ship routing and scheduling has blossomed during the last decade. Comparing to the former decade its volume has more than doubled, and the same is true for the variety of research outlets. The research seems to be catching up with the increasing world fleet and trade. Problems of wider scope have been addressed, specifically liner network design, maritime inventory routing, and maritime supply chains [13].

Ship routing has a major role in fuel consumption and performance of vessels, nowadays, fortunately, autonomy and total energy efficiency are increasingly close, there are already several companies working in this direction.

One example that proves the course of this trend is Yara Birkeland, which will be the world's first fully electric and autonomous container ship, with zero emissions. With this vessel, Yara will reduce diesel-powered truck haulage by 40,000 journeys a year. It scheduled to set sail in 2020 (**Figure 5**).

According to Secretary General Kitack Lim (2017), "The seven strategic directions point us now towards more effective rule-making and implementation processes by integrating new and advancing technology to respond to our challenges, among others, to increase ship safety, including addressing new emerging technologies such as autonomous vessels".

This leads to new challenges and some controversy, but we need to move forward so the use of hi-tech is the answer to achieve the excellence at the rate that environment needs.

3.5 Hi-tech in ship design

Building on the basis of cyber-physical production systems (CPPS) which merges the real and virtual space ("Industry 4.0: Challenges and solutions for the digital transformation and use of exponential technologies," [14]).

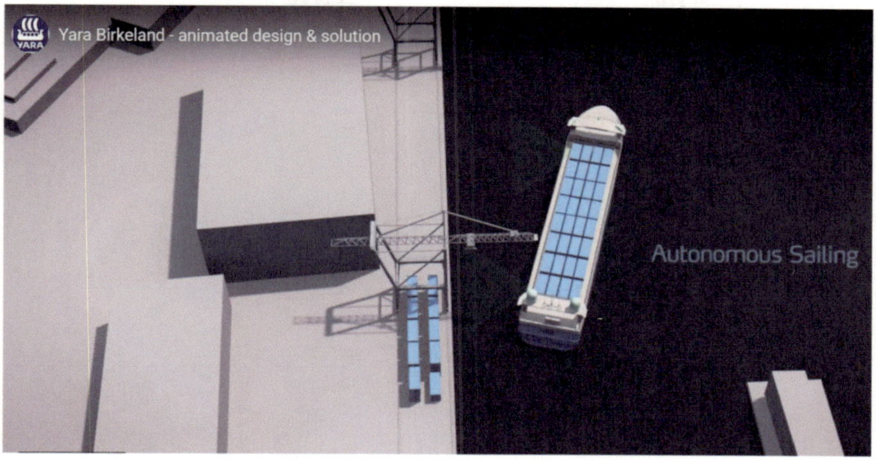

Figure 5.
Typical pollution sources from vessels. Source: Yara International ASA.

From the conception, to the construction and operational fazes we must use and enjoy the best that technological advances have to give, to achieve excellence results faster and more detailed.

One of the major evolutions in designing process was without doubt the use of CAD (Computer Assisted Drawing). CAutoD, commonly known as virtual rapid prototyping is an extension of traditional Computer-Aided Design (CAD) [15].

According to Ang et al. [10]. It makes use of biologically inspired machine learning techniques such as an evolutionary algorithm (EA) to intelligently search and evaluate the design space for innovative and optimal solutions. Coupled with powerful evaluation tools, this forms a close loop to fully automate the design process.

Nowadays, designers can count on the work excellence and speed that BIM (Building Information Models) has come to bring. Building Information Modeling (BIM) is a process used by Architecture Engineering Construction (AEC) stakeholders, which simulates a construction project in a multi-dimensional digital model and provides multitudes of project benefits from project inception to its occupancy [16]. This is without doubt the path to follow to achieve faster and better projects, much easier to build and to adapt in case of necessity. Other powerful tool is 3D Laser Scanning & Reverse Engineering. Notably, 3D laser scanning techniques can provide an accurate surface of the tested elements consisting of point clouds, which can reflect the actual spatial performance of the element. In addition, point clouds can be transformed into actual models and compared with design models through reverse modeling [17].

Figures 6–9 make a comparison between a project in the past century and these days (2018).

Figure 6.
Hand-made original blueprint of titanic provided by ultimate titanic.

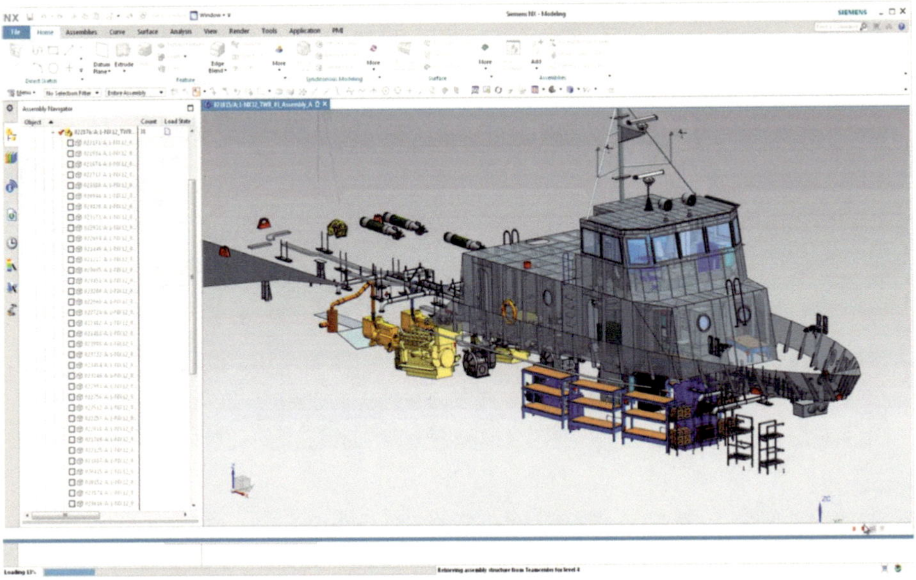

Figure 7.
BIM based ship design in NX-12 software by CAMdivision.

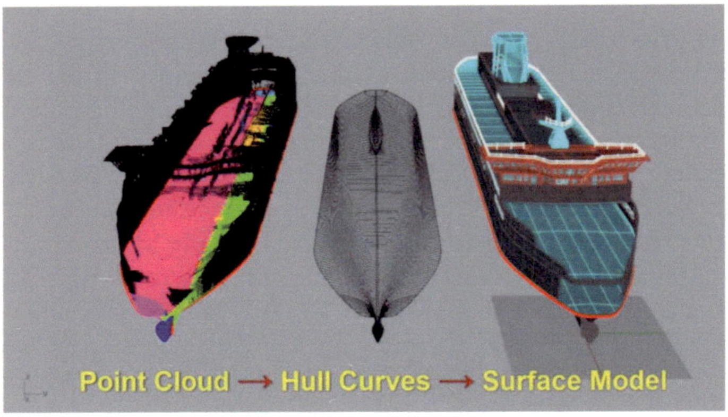

Figure 8.
3D laser scan & point cloud & reverse engineering by summit 3D.

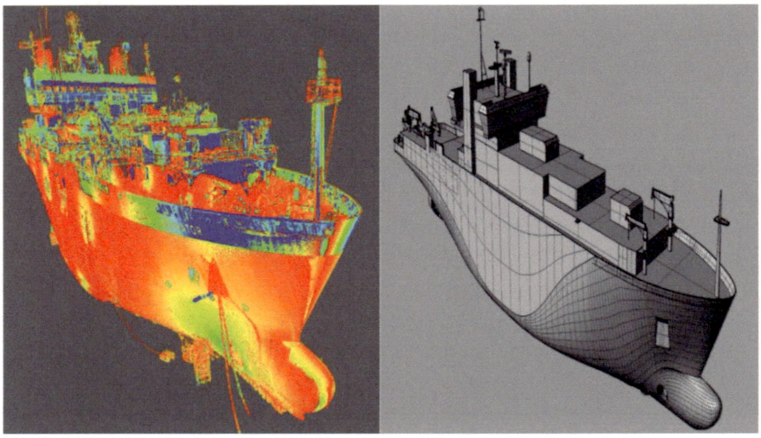

Figure 9.
3D laser scan & reverse engineering by SEVERNPARTNERSHIP.

4. Conclusions

In light of the above, it is possible to understand the importance of vessels and to aware, all the stakeholders involved in navigation, transport and design of ships, to work around a common goal: boosting performance and efficiency, ensuring the preservation of the environment.

Investigations must be carefully accessed before developing and launching products on the market. Simultaneously, the speed demanded is increasing, because the novelty of today becomes obsolete in little time.

Of course, these innovations are costly for businesses, never the less there's much to gain from them in a long term.

In addition, policy variables reflecting regulations at seaports affect port efficiency in a nonlinear way, its paramount to implement a global policy towards sustainability, regulating in very brief way the type of fuel used, CO_2 emissions, oil spills, the coatings used, waste waters and cargo residues discharges. Shipping design has to become smarter and more eco-friendly in order to increase vessels efficiency and to reduce their environmental footprint.

However, it is not enough to create rules behind rules, despite the excellent commitment of some governments and organizations, the most important guideline is to continue to raise awareness of companies and people more connected to this industry reprograming their way of acting and thinking. With this, it becomes possible achieving a utopia, in which the ships have zero emissions, their inks do not pollute, their maintenance is done with the proper periodicity, thus ceasing (partially) with the nefarious effect that our species has been provoking on the planet.

It should be noted that this type of measures will have to be implemented in the shipbuilding industry as well as in all industries.

As a final note, we will have to face the storms that lobbys enemies of the environment will cause and paddle against the tide of global pollution in order to achieve a better world.

Author details

Sérgio António Neves Lousada[1,2,3], João Pedro Gouveia[1]
and Rui Alexandre Castanho[2,3,4,5,6]*

1 Department of Civil Engineering and Geology, Faculty of Exact Sciences and Engineering, University of Madeira, Funchal, Portugal

2 VALORIZA - Research Centre for Endogenous Resource Valorization, Portalegre, Portugal

3 Institute of Research on Territorial Governance and Inter-Organizational Cooperation, Poland

4 Environmental Resources Analysis Research Group (ARAM), University of Extremadura, Badajoz, Spain

5 ICAAM—Institute for Agrarian and Environmental Sciences, Évora, Portugal

6 Faculty of Applied Sciences, WSB University, Poland

*Address all correspondence to: acastanho@wsb.edu.pl

References

[1] Pearlman H. How Much Water Is There On, In, and Above the Earth? USGS Water Science School; 2016

[2] Khyade VB, Mankombu SS. Water: The Pacemaker for Life of Earth; 2016

[3] Soehngen et al. PIANIC Report N°99—Considerations to Reduce Environmental Impact of Vessels; 2008

[4] Zhang F, Sun XX, Zhou Y, Zhao CJ, Du ZH, Liu RY. Ecosystem health assessment in coastal waters by considering spatio-temporal variations with intense anthropogenic disturbance. Environmental Modelling & Software. 2017;**96**:128-139. DOI: 10.1016/j.envsoft.2017.06.052

[5] Hua Y, Zhua D. Empirical Analysis of the Worldwide Maritime Transportation Network; 2008

[6] UNCTAD (United Nations Conference on Trade and Development). Review of Maritime Transport; 2012

[7] UNCTAD (United Nations Conference on Trade and Development). Review of Maritime Transport; 2018

[8] Park SH, Inwon L. Optimization of drag reduction effect of air lubrication for a tanker model. International Journal of Naval Architecture and Ocean Engineering. 2018;**10**:427-438

[9] Corbett JJ, Wang H, Winebrake JJ. The effectiveness and costs of speedreductions on emissions from international shipping. Transportation Research Part D: Transport and Environment. 2009;**14**(December (8)):593-598. DOI: 10.1016/j.trd.2009.08.005. ISSN 1361-9209

[10] Ang JH, Goh C, Li Y. Smart Design for Ships in a Smart Product Through-Life and Industry 4.0 Environment; 2016

[11] Ejenstam L et al. The effect of superhydrophobic wetting state on corrosion protection-the AKD example. Journal of Colloid and Interface Science. 2013;**412**:56-64

[12] Chatterjee U, Bose SK, Roy SK. Environmental Degradation of Metals: Corrosion Technology Series/14. CRC Press; 2001

[13] Christiansen M, Fagerholt K, Nygreen B, Ronen D. Ship routing and scheduling in the new millennium. European Journal of Operational Research. 2013;**228**:467-483

[14] Deloitte Report. Industry 4.0: Challenges and solutions for the digital transformation and use of exponential technologies; 2014

[15] Li Y et al. CAutoCSD, evolutionary search and optimization enabled computer automated control system design. International Journal of Automation and Computing. 2004;**1**(1):76-88

[16] Fountain J, Langar S. Building information modeling (BIM) outsourcing among general contractors. Automation in Construction. 2018;**95**:107-117

[17] Liu J, Zhang Q, Wu J, Zhao Y. Dimensional accuracy and structural performance assessment of spatialstructure components using 3D laser scanning. Automation in Construction. 2018;**96**:324-336